OXFORD CHEMISTRY PRIMERS

Physical Chemistry Editor	Founding/Organic Editor	Inorganic Chemistry Editor	Chemical Engineering Editor
RICHARD G. COMPTON	STEPHEN G. DAVIES	JOHN EVANS	LYNN F. GLADDEN
University of Oxford	University of Oxford	University of Southampton	University of Cambridge

Organic Stereochemistry

Michael J. T. Robinson

The Dyson Perrins Laboratory and Magdalen College, University of Oxford

OXFORD

UNIVERSITY PRESS

This book has been printed digitally and produced in a standard specification
in order to ensure its continuing availability

OXFORD
UNIVERSITY PRESS

Great Clarendon Street, Oxford OX2 6DP

Oxford University Press is a department of the University of Oxford.
It furthers the University's objective of excellence in research, scholarship,
and education by publishing worldwide in

Oxford New York

Auckland Cape Town Dar es Salaam Hong Kong Karachi
Kuala Lumpur Madrid Melbourne Mexico City Nairobi
New Delhi Shanghai Taipei Toronto
With offices in
Argentina Austria Brazil Chile Czech Republic France Greece
Guatemala Hungary Italy Japan South Korea Poland Portugal
Singapore Switzerland Thailand Turkey Ukraine Vietnam

Oxford is a registered trade mark of Oxford University Press
in the UK and in certain other countries

Published in the United States
by Inc., New York

© Michael J. T. Robinson 2000

ISBN 0-19-879275-1

Antony Rowe Ltd., Eastbourne

Series Editor's Foreword

An understanding of stereochemistry, the three-dimensional properties and behaviour of molecules, is an intrinsic part of all organic chemistry. The ability to think in three dimensions is a prerequisite to being able to understand and utilise the basic principles of organic chemistry. It is not surprising therefore that stereochemistry plays a vital role as a fundamental topic throughout all undergraduate organic chemistry courses.

Oxford Chemistry Primers have been designed to provide concise introductions relevant to all students of chemistry and contain only the essential material that would be covered in an 8–10 lecture course. In the present Primer Mike Robinson presents the basic concepts of stereochemistry in a logical and easy to follow fashion. This Primer provides essential reading for apprentice and master chemist alike.

Professor Stephen G. Davies
The Dyson Perrins Laboratory, University of Oxford

Preface

Some aspects of organic stereochemistry have been covered in other books in the Oxford Chemistry Primers series, notably in L. M. Harwood's *Polar Rearrangements* and A. J. Kirby's *Stereoelectronic Effects* and are not repeated here. I have also reluctantly left out the stereochemistry of natural and synthetic polymers, supramolecular chemistry, photochemistry, and the origin of the chiral bias in biological chemistry.

I wish to thank Drs Carolyn Carr and Gordon Whitham for extensive and very helpful criticisms of the draft manuscript and Professor Steve Davies for patience beyond the call of duty for an editor.

Oxford   M. J. T. R.
May, 1999

Contents

1 Introduction: development of methods and concepts in organic stereochemistry

Understanding of organic stereochemistry benefits, more than most branches of chemistry, from a knowledge of how it has developed. This chapter is a very brief survey of the history of organic stereochemistry to about 1960 and introduces many of the more important terms used in organic stereochemistry.

1.1 Before van't Hoff and Le Bel

The first experimental evidence that the shapes of molecules might be significant came from the observation of *optical activity* (rotation of the plane of polarization of plane polarized light) in organic liquids (Biot, 1815). This was taken by Fresnel (1824) to imply that such molecules are helical but this was not followed up.

Pasteur's recognition of *enantiomerism* in tartaric acids was hugely influential. At the beginning of his work three 'tartaric acids' were known. He showed that *racemic* tartaric acid was a 1 : 1 mixture of the known (+)-form and a new (−)-form (1848). These two forms of tartaric acid were indistinguishable by most chemical and physical criteria but were related as an object to a non-identical mirror image. For example, these acids had optical rotations of *equal magnitude but opposite signs* in solution and the crystal habits of some of their salts were related as a left hand is to a right hand. He correctly inferred that because the differences persisted in solution the *molecules* of these two tartaric acids must have enantiomeric structures, like a pair of hands. He also showed how a pair of enantiomeric compounds could be separated and that they underwent *biochemical* reactions at different rates. Pasteur described the optically active tartaric acids as *dissymmetrique* (meaning lacking reflection symmetry) which unfortunately was translated as *asymmetric* (having *no* symmetry). The confusion has been removed by using the term chiral (derived by Kelvin from the Greek word χειρ, *cheir*, for hand).

The understanding of these stereochemical differences required an understanding of (i) the *constitutions*, i.e., the connections between atoms, of organic molecules and (ii) the geometry of carbon atoms. The first required the recognition of the tetravalency of carbon (Kekulé, 1858). The second came from van't Hoff and Le Bel.

> Enantiomeric molecules are related as an object to its *non-identical* image in a plane mirror. (+)- and (−)-, from the sign of optical rotation of a compound under standard conditions, are used as prefixes to chemical names of enantiomeric forms of a compound showing optical activity.
> The first three forms of tartaric acid to be isolated were the (+)-form from wine fermentation and two optically inactive forms, *racemic* and *meso*. Racemic is now used as an adjective for all 1 : 1 mixtures of enantiomeric compounds. *Meso* is similarly used for *achiral* compounds containing compensating pairs of chiral centres or other chiral elements (see pages 45-47, 51).

1.2 The tetrahedral carbon atom

Van't Hoff and Le Bel in 1874 independently explained the existence of pairs of enantiomeric molecules such as (+)- and (−)-tartaric acids and other newer examples, notably the two lactic acids from milk and from muscle. They supposed that the four valencies of a carbon atom were directed towards the

> The first reference to a tetrahedral carbon atom was made by Paternò (see page 3).

Van't Hoff's work prompted a famous and ill-judged polemic by Kolbe that served to draw attention to the younger chemist's work.

H   H   H CO_2H
 \   /   \   /
  C=C   C=C
 /   \   /   \
CO_2H CO_2H CO_2H H

Maleic acid   Fumaric acid

Van't Hoff's and Le Bel's work on stereochemistry came before Nobel prizes were instituted but van't Hoff was later awarded a Nobel prize for his work in *physical* chemistry.

four apices of a regular tetrahedron with carbon at the centre and that there is free rotation about single bonds. Four *different ligands* (atoms or groups) can be attached to such a carbon atom in *two* distinguishable ways to form enantiomeric molecules. Van't Hoff developed this model further. He showed that if two such carbon atoms were joined by a *double bond*, i.e., were joined through two pairs of tetrahedral apices forming two *bent* bonds, there should be no rotation about such a bond and with suitable substituents *geometrical isomerism* should result. He used this to explain the previously puzzling relationships between *maleic* and *fumaric* acids and similar 'geometrical isomers' (now called *diastereomers*).

One of van't Hoff's predictions, that certain disubstituted *allenes*, molecules with the unit C=C=C, should exist in enantiomeric forms, was not verified until 1935! His assumption of free rotation about single bonds was shown to be misleadingly incomplete at about the same time (see Section 1.5).

The idea of the tetrahedral carbon atom led to studies of stereoisomerism in molecules with tri- and tetracoordinate atoms such as silicon, phosphorus, and sulfur until such methods were made obsolete by physical methods for determining the shapes of molecules (pages 3, 4).

1.3 Relative configurations

CO_2H CO_2H
  | |
H—+—OH   HO—+—H
  | |
HO—⋯⋯H   H⋯⋯—OH
  | |
CO_2H CO_2H

   1.1   1.2

A tetra-coordinate tetrahedral atom is *chiral* if a distinguishable molecule results from the exchange of *any* two ligands. A right hand is the non-identical mirror image of a left hand.

When van't Hoff and Le Bel explained the structures of the enantiomeric tartaric acids they had no way of telling which acid was which, i.e., whether the molecules of (+)-tartaric acid matched the three dimensional structure **1.1** or its mirror image **1.2**: note that rotation about the single bonds does *not* interconvert **1.1** and **1.2**. This problem was intriguing but its solution was not urgent because enantiomeric molecules differ only in properties that change on reflection, e.g., optical rotation. Many chemists thought that this problem would never be solved and a general solution came only in 1951 (page 3).

Stereoisomeric molecules with two or more chiral carbon atoms, however, presented a much more pressing problem. Two such molecules are *either* enantiomeric *or diastereomeric*. Diastereomeric molecules differ in *all* properties that depend on intramolecular distances, e.g., heat of formation, and it was necessary to know *relative* configurations, i.e., the relative arrangements of ligands around two or more chiral atoms, if properties and reactions were to be correlated with stereochemical structure.

CHO   CHO
 | |
H—C—OH   HO—C—H
 | |
HO—C—H   H—C—OH
 | |
H—C—OH   HO—C—H
 | |
H—C—OH   HO—C—H
 | |
CH_2OH CH_2OH

   1.3   1.4

Simple sugars (polyhydroxy-aldehydes and -ketones) provided the first major problem in determining relative configurations. This was solved by E. Fischer who was able to show that, e.g., the structure of the molecules of (+)-glucose is *either* **1.3** *or* **1.4** but not a member of one of the other seven pairs of enantiomers with the same constitution. With no prospect at the time of solving the problem of absolute configuration Rosanoff (1906) suggested that the simplest chiral sugar, (+)-glyceraldehyde **1.5**, should *arbitrarily* be assigned the configuration shown and that the configuration of other chiral molecules should be chemically correlated with that of (+)-glyceraldehyde.

CHO
 |
H—C—OH
 |
CH_2OH

   1.5

A limitation of Fischer's methods for determining relative configurations was that he had no way of determining whether or not reactions at chiral centres could change the arrangement (retention or inversion of 'configuration': a different use of configuration, see pages 3, 66) of groups on such an atom. This required studies of *reaction mechanisms*.

1.4   Stereochemistry and reaction mechanisms

Chemists originally assumed that a simple substitution R–X → R–Y would take place without any change in R, i.e., with *retention of configuration* at R if the reaction site was a chiral carbon atom. Some reaction sequences that converted chiral compounds into their enantiomers (Walden, 1896) showed that *inversion* of configuration must occur in some substitutions but failed to show which. In yet other substitutions *chiral* compounds formed *racemic* products. These problems were solved only by applying quantitative methods to the study of organic reaction mechanisms.

Simple *nucleophilic* substitutions were shown to take place in *either* a single step bimolecular mechanism (S_N2), invariably with *inversion*, *or* a two step reaction with an initial unimolecular step (S_N1) with more or less racemization (Ingold, 1927-1930). *Retention* was shown to be a less common result of a more complex mechanism. This was followed by studies of the stereochemistry of additions, eliminations, and rearrangememts, leading finally to an understanding of *pericyclic* reactions (Woodward and Hoffmann, 1969). The stereochemical course of most reactions is now predictable.

Retention of configuration

Inversion of configuration

1.5   Conformational analysis

The first reference (Paternò, 1869) to the possible significance of internal rotation about the a C–C single bond, that in 1,2-dichloroethane, *preceded* van't Hoff's and Le Bell's work but was not followed up. In 1936 Pitzer found a small difference between experimental and computed entropies of ethane and showed that this implied that internal rotation about the C–C bond is not free but has a small barrier. Such barriers are too small to allow separable isomers in derivatives such as 1,2-dichloroethane but force most molecules into a limited number of *conformers*, detectable by diffraction and spectroscopic methods (see Section 1.6), corresponding to the conformations close to potential energy minima for such internal rotations.

The relative rotation of the groups at the two ends of a single bonds is called *internal rotation*. The (infinite number of) arrangements that result are *conformations*.

Sachse (1890) showed that six tetrahedral carbon atoms could form two non-planar forms of cyclohexane. Because the number of isomers of derivatives of cyclohexane could be explained by a *planar* ring, Sachse's idea was neglected until Mohr (1918) used it to predict two isomers of 'decalin', which were isolated by Hückel (1925). The general significance of non-planar rings was largely ignored until Hassell (1943) showed by electron diffraction that cyclohexane and *cis-* and *trans-*decalin existed in chair *conformers*. Soon Barton (1950) demonstrated the relationship between conformation and reactivity in complex compounds and founded *conformational analysis*.

Cis- and *trans-*decalin: the 'chair' shaped rings were established by electron diffraction.

Barton and Hassell shared the Nobel Prize for chemistry in 1969.

1.6   X-ray diffraction: absolute configuration

The discovery of X-ray diffraction in simple inorganic crystals (von Laue, 1913) in time led to studies of organic molecules: the first was the *planar* hexamethylbenzene (Lonsdale, 1929). Later studies confirmed the qualitative reliability of van't Hoff's predictions but refined them by determining bond lengths and bond angles. The growing power of the method for stereochemistry can be gauged from some landmark structures and dates:

cholesteryl iodide: *relative* configuration (8 chiral centres; Crowfoot, 1945);
sodium rubidium tartrate: first *absolute* configuration (Bijvoet, 1951);
double helix in DNA (Watson, Crick, Wilkins, and Franklin, 1953);
tertiary structure of haemoglobin (Perutz, 1970)

A neutron diffraction experiment analogous to Bijvoet's with X-rays allowed the absolute configuration of $HOCHDCO_2^-Li^+$ to be determined (1965).

The determination of the absolute configuration of (+)-tartaric acid and therefore the absolute configurations of the many chiral compounds which had been related to it by chemical correlations resolved the intellectual problem left by Fischer and others. It is now almost routine to determine the absolute configuration of molecules in chiral crystals studied by X-ray diffraction and there are very reliable *networks* of evidence establishing the absolute configurations of most chiral molecules beyond doubt.

1.7   Electron diffraction and spectroscopic methods

X-ray diffraction has enormous power to determine structure but the crystalline state normally allows only one conformation of a molecule to be present in a crystal. This limitation is removed by physical methods that may be applied to fluid phases.

The preferred conformations of
1,2-dichloroethane

Electron diffraction (gas phase) and vibrational spectroscopy (all phases) are mostly limited to simple molecules so far as stereochemistry is concerned. The establishment of the preferred conformations of 1,2-dichloroethane (1935) showed how close Paternò had been in 1869!

Pure rotation (microwave) spectra are even more limited than vibrational spectroscopy in terms of the complexity of molecule that can be studied fully. However, they give invaluable and very detailed information about bond lengths, bond angles, conformational equilibria, shapes and heights of barriers to internal rotation, and dipole moments and their directions within molecules.

Nuclear magnetic resonance spectroscopy holds a unique place in stereochemistry in general and conformational analysis in particular because there are such a great variety of experiments possible, many yielding information not otherwise available, e.g.:

- The rate of ring chair to chair inversion in cyclohexane (Jensen, 1960: note that this is a degenerate process that leads to no observable change in other spectroscopies or in diffraction experiments).

- Peptide and protein folding in solution.

- Enzyme-substrate complexes in solution.

1.8   And so on

Organic stereochemistry is still developing rapidly and neither this short Introduction nor the rest of the book can cover the whole subject. In particular I have been forced to leave out the stereochemical aspects of polymers (synthetic and natural), supramolecular chemistry, and photochemistry, and the origin of the chiral bias in biological chemistry.

Further reading

W.A. Bonner, "Origins of Chiral Homogeneity in Nature", *Top. Stereochem.*, 1988, **18**, 1.

M. Farina, "The Stereochemistry of Linear Macromolecules", *Top. Stereochem.*, 1987, **17**, 1.

J.-M. Lehn, *Supramolecular Chemistry*, Wiley, New York, 1995.

S.F. Mason, "The Foundations of Classical Stereochemistry", *Top. Stereochem.*, 1976, **9**, 1.

2   Organic molecules: shapes, sizes, and strain

Organic stereochemistry is the study of the relationships between three dimensional structure and properties of organic molecules and compounds. This chapter begins with a *description* of the 'shapes' of simple organic molecules, and the parameters (bond lengths and angles, torsion angles, and van der Waals radii) used to characterize them. The chapter continues with methods for representing three dimensional structures in two dimensions, which can be supplemented by molecular models and, increasingly, computer graphics. This is followed by an account of *explanations* of such shapes, which provides a framework for describing the 'shapes' of C atoms (see below) in transient species and of heteroatoms in organic compounds. The symmetry of finite molecules is briefly described with special emphasis on chiral and achiral molecules. The chapter ends with an account of 'strained' organic molecules.

The Valence Shell Electron Pair Repulsion (VSEPR) explanation of the shapes of molecules is not very useful in organic chemistry and will not be covered here.

2.1   Structures of organic molecules

When you began to study organic chemistry you probably limited your interest to the *constitutions* of molecules, i.e., the atoms and the bonds connecting them. You soon learnt to condense these formulae, to save time and to concentrate attention on functional groups, e.g., in propanal (propionaldehyde) it is often the aldehyde group that is most important (Fig. 2.1). Such formulae allow us to follow simple chemistry but are limited for mechanisms and useless for stereochemistry, which requires drawings (pages 7–9), molecular models or computers graphics (page 9) to bring out the 3D *structures* of molecules. Use *structure* to include stereochemical aspects of molecules.

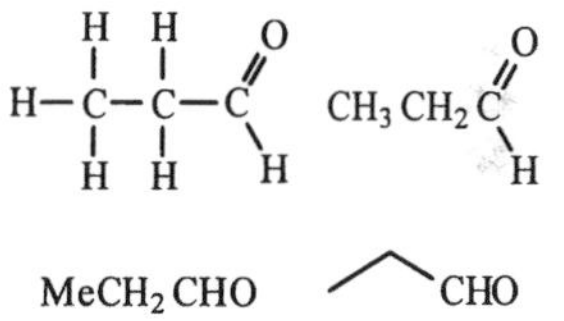

Fig. 2.1. Four formulae for propanal (propionaldehyde).

2.2   Shapes of simple organic molecules: bond lengths and bond angles

The largest source of information about the shapes and sizes of molecules comes from *single crystal* X-ray diffraction. Electron diffraction and microwave spectroscopy using *gases* provides very accurate information for simple molecules, while other techniques are useful in special cases.

The simple structures of propanal shown above basically show the topology or connectivity of propanal. More information about propanal can be given if a structure includes the *shape* of the molecule. This requires four types of parameter: bond lengths, bond angles, torsion angles, and van der Waals radii.

A bond length is the distance between the nuclei of a pair of bonded atoms. It is fairly constant from molecule to molecule for a given bond type, i.e., for a given pair of atoms and multiplicity of bonding. Quite small deviations from 'normal' values can be chemically significant. Bond lengths are typically 100 to 250 pm (Fig. 2.2), with an accuracy of about ±1 pm for ordinary quality measurements.

All molecules vibrate, even at 0 K. A molecular 'shape' is therefore not a simple idea, e.g., bond lengths measured by different methods may differ by ~1 pm .

Units for bond lengths:

1 nm	$= 1 \times 10^{-9}$ m
1 Å (Ångstrom)	$= 1 \times 10^{-10}$ m
1 pm	$= 1 \times 10^{-12}$ m

This book will use pm.

H₃C—H	H₃C—CH₃	H₃C—NH₂	H₃C—OH	H₃C—F	H₃C—Cl	H₃C—Br	H₃C—I
109	154	148	142	139	178	194	214

| Bond lengths *decrease* across a period | | Bond lengths *increase* down a group |

Fig. 2.2. Lengths/pm of single bonds between carbon and another atom in saturated molecules

The length of a carbon-carbon bond comes within a small range, about 120 to 154 pm in ordinary molecules (Fig. 2.3). Its value mainly depends on the multiplicity of the bond and of the other bonds formed by the two C atoms.

H₃C—CH₃	H₃C—CH=	H₃C—C≡	=HC—CH=	=HC—C≡	≡C—C≡	H₂C=CH₂	H₂C=C=	=C=C=	HC≡CH	Benzene
154	150	146	148	143	138	134	131	128	120	139
sp^3–sp^3	sp^3–sp^2	sp^3–sp	sp^2–sp^2	sp^2–sp	sp–sp	sp^2–sp^2	sp^2–sp	sp–sp	sp–sp	sp^2–sp^2
	Single						Double		Triple	'1.5'

Fig. 2.3. Examples of carbon carbon bond lengths/pm and their relation to the hybridization of the C atoms (see Section 2.7)

It is common to specify the shape of a carbon atom by reference to the relative directions of the bonds to its ligands. *Experimentally* the shapes of tetravalent carbon atoms correlate primarily with the number of ligands. *Explanations* of this will be outlined later (pages 9–12).

C atoms can have five or six ligands but this does *not* mean five or six 'bonds': see Section 2.8.

A carbon atom is described as 'linear' or 'digonal' or '*sp*' (two ligands), 'planar' or 'trigonal' or '*sp²*' (three ligands), or 'tetrahedral' or '*sp³*' (four ligands). For some purposes, e.g., in deciding the numbers and types of isomers of a given molecular formula, idealized bond angles, related to the number of ligands, suffice. The multiplicity of the bonds to linear and planar carbon atoms is unimportant at a qualitative level (Fig. 2.4).

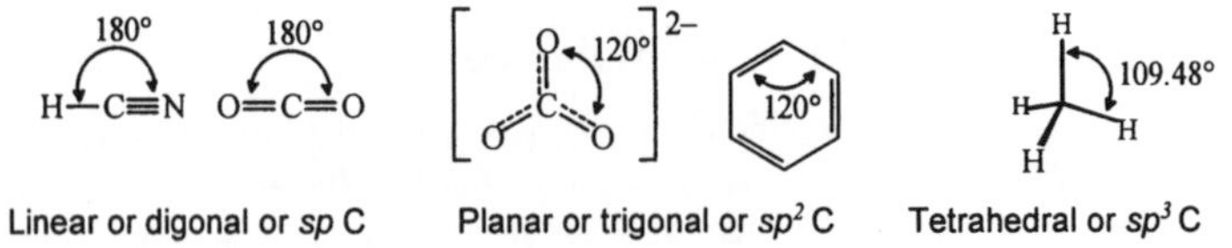

Linear or digonal or *sp* C   Planar or trigonal or *sp²* C   Tetrahedral or *sp³* C

Fig. 2.4. Ideal bond angles in very simple molecules and ions.

The molecules in Fig. 2.4 are very symmetrical, unlike most organic molecules, and most bond angles differ from these ideal values (see Fig. 2.5).

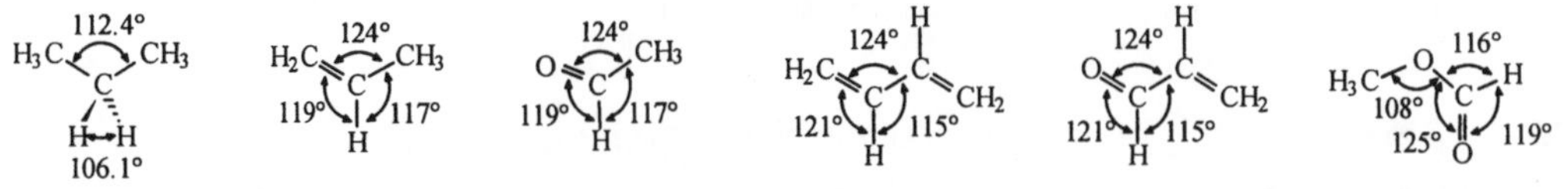

Fig. 2.5. Examples of bond angles in simple unsymmetrical, acyclic molecules with little or no strain

Bond angles are more easily deformed than bond lengths (see pages 17–19). Bond angles in cyclic molecules and in strained molecules may be very different from those in simple acyclic molecules.

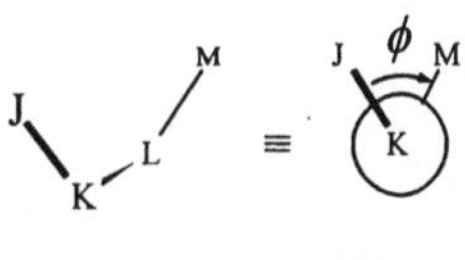

Fig. 2.6. A *positive* torsion angle ϕ. N.B. It does not matter from which end of K–L the latter is viewed.

2.3 Torsion angles

In a chain of atoms J–K–L–M (neither J–K–L nor K–L–M collinear) the torsion angle ϕ for the bonds J–K and L–M is the angle between the *projections* of J–K and L–M on a plane perpendicular to K–L: ϕ is defined as *positive* if the *front bond must be rotated* (≤180°) *clockwise* about K–L to make the projections of K–J and L–M coincide (Fig. 2.6), otherwise it is *negative*. *Internal rotation* changes the torsion angle ϕ.

There are usually two or more values of a torsion angle ϕ that correspond to potential energy minima. Internal rotation about a single bond is usually easy and is sometimes misleadingly referred to as 'free rotation'.

In simple molecules the preferred torsion angle for a pair of bonded C atoms depends on the number of ligands on the C atoms (Fig. 2.7).

> A torsion angle is *not* the same as a dihedral angle, which is the unsigned angle ($0 \leq \phi \leq 180°$) between two planes meeting at a common edge.

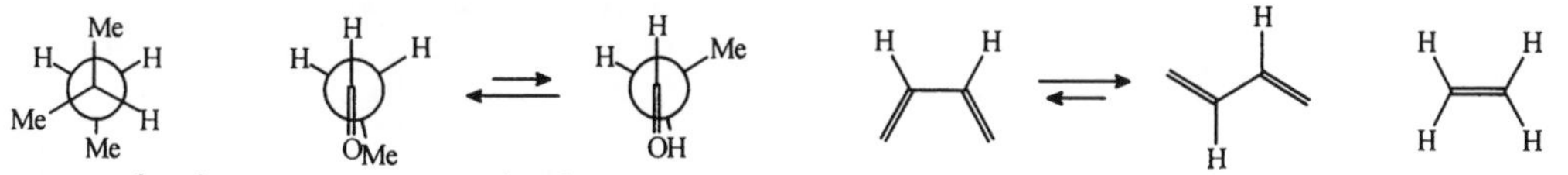

C atoms: sp^3–sp^3
ϕ ~±60° or 180°

sp^3–sp^2
ϕ = 0° or ~120° (Me**C**–C=O)
~±60° or 180° (Me**C**–CH)

sp^2–sp^2: C–C
ϕ 0° (*s-cis*)* or 180° (*s-trans*)*

sp^2–sp^2: C=C
ϕ 0 or 180°

Fig. 2.7. Torsion angles for potential energy minima for CC single and double bonds in some simple molecules.
*Prefixes *s-cis* and *s-trans* are often used for conformers differing in torsion angles for sp^2–sp^2 single bonds.

2.4   Van der Waals' radii and 'non-bonded interactions'

Van der Waals' radii were originally a way of characterizing the volumes of gaseous molecules, treating them as hard spheres like pool balls. In reality molecules are 'squashy'. Van der Waals' radii now refer to the distances between specified atoms or groups (in the same or different molecules) at which there is no net attraction or repulsion (see Fig. 2.8).

Adding imaginary spherical shells, with radii equal to the van der Waals' radii (Fig. 2.9), to atoms or groups at the outside of a molecule gives an idea of the 'size' of that molecule. Fig. 2.10 illustrates the space molecules occupy ('sticks and dots') compared with more usual representations.

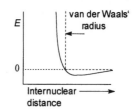

Fig. 2.8. The potential energy E (0 at ∞) of interaction of two molecules

H	N	O	S	F	Cl	Br	I	CH$_3$
125	150	140	185	135	180	195	215	200

Fig. 2.9. Approximate values of van der Waals' radii/pm

Fig. 2.10. Three ways of representing 4-t-butylcyclohexanone

Better results are achieved by a combination of attractive and repulsive *van der Waals forces* in molecular mechanics calculations (Section 3.14).

2.5   Conventions used in drawing organic molecules

Graphic formulae representing the stereochemistry of molecules and compounds serve two distinct purposes. The drawings in this section are simplifications of accurate perspective diagrams. They give useful ideas of the relative positions of atomic nuclei and sometimes an idea of the space occupied by atoms or groups, *at a given instant*. Remember that real molecules are in constant motion. The other type of stereochemical formulae are conventional ways of indicating the *stereoisomerism of compounds* (Section 4.6).

Three ways of viewing a bond and the associated atoms

N.B. Most chemical structures in books are *not* accurate. They are based on *stylised* shapes to allow simple units to be joined together by a chemical drawing program.

Often two atoms and the bond between them provide the centre of interest. The bond may be viewed in three ways using ethane as a simple example (Fig. 2.11). Ethane is drawn in the lowest energy staggered conformation in the first row and in the eclipsed conformation in the second.

C–C sideways on   C–C oblique ('sawhorse')   C–C end-on (Newman)

In the 'Newman' projection for the eclipsed conformation the methyl group at the front is rotated a little to avoid completely obscuring that at the back.

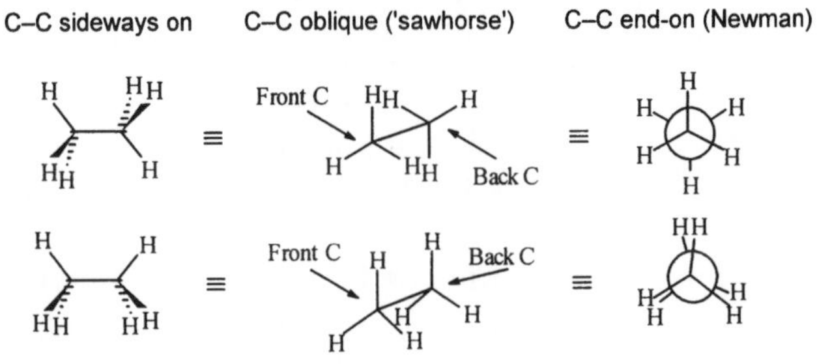

Fig. 2.11.

The Newman projection is my favourite for depicting acyclic molecules but you should find out which you like best for your own studies.

These conventions may be combined. The usual way of drawing the chair conformation of cyclohexane combines four 'sawhorse' C–C bonds and two side views of C–C bonds (Fig. 2.12), usually with most of the exocyclic bonds omitted.

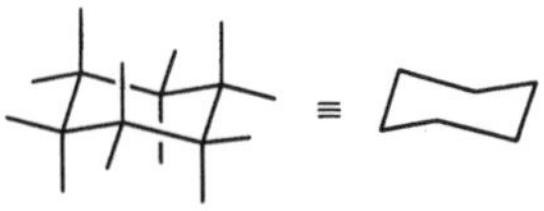

(i)   (ii)
Fig. 2.12. Cyclohexane shown (i) with and (ii) without the C–H bonds

How to draw the chair conformation of cyclohexane

There is no 'correct' way to do this and I describe the method I prefer. First draw a 'V' rotated by ~45° from the vertical (a): these C–C bonds are foreshortened (Fig. 2.13). Add two *parallel* sloping lines (b, c: *not* foreshortened), and then two lines (d, e), each of which should be parallel to one or the other of the two lines in (a): opposite pairs of sides are parallel. Often it is sufficient to add a small number of exocyclic bonds: Fig. 2.13 (f) to (i) covers the addition of all twelve. The six axial bonds (*a*) are vertical and parallel to the *three-fold* symmetry axis C_3 of the ring (f) (assuming ideal bond angles). Pairs of equatorial bonds (*e*) at opposite ends of the ring (g, h, and i) are parallel to one another and to a pair of sides of the ring. *None* of the bonds are horizontal and the equatorial bonds slope up and down alternately.

An ability to draw the chair conformations of cyclohexane and related compounds with reasonable accuracy is very helpful in understanding their stereochemistry.

How *not* to draw cyclohexane! Two examples from textbooks.

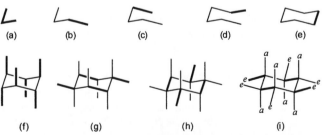

(a)   (b)   (c)   (d)   (e)

(f)   (g)   (h)   (i)
Fig. 2.13. Steps in drawing cyclohexane in the chair conformation

The axial C–H bonds in the real molecule (∠CCC = 111.4°) lean outwards slightly.

The most obvious way to draw even a *planar* molecule is often not the best way to bring out stereochemical features of reactions. For example, an alkene is usually viewed from one *face* but this is a poor choice for depicting addition reactions of alkenes. The latter are better seen in a perspective from one *side* or from one *end* (Fig. 2.14).

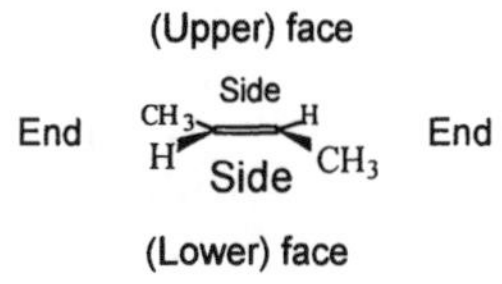

(Upper) face

End   CH₃―Side―H   End
H―Side―CH₃

(Lower) face

Fig. 2.14. Names of regions of space around C=C in an alkene

2.6   Molecular models and computer graphics

Simple molecular models fall into two main classes:

Skeletal models: These models show the relative positions of atomic nuclei in much the same way as drawings. They vary from cheap 'ball and stick' models, which you should buy for yourself or have access to, to the precision Dreiding stereomodels.® Such models help in understanding stereoisomerism and reaction mechanisms. They may have quite accurate bond angles and relative bond lengths but do not represent overcrowded molecules well.

Remember that even the best mechanical models are static, unlike molecules.

Double bonds and aromatic compounds are handled in three different ways. Dreiding models use relatively large molecular fragments such as ethene or benzene as flat rigid units which directly represent only the σ-framework (see following Section). Some models include a π-electron framework to give more realistic relative dimensions to unsaturated molecules. In the cheapest 'ball and stick' models double bonds are formed from two *bent* single bonds, much in the way suggested by van't Hoff's model of a double bond (page 2).

Space filling models: When we are interested in how a molecule fits in contact with another molecule we can use 'space filling' models. These take account of the van der Waals' radii of atoms but are expensive, troublesome to handle, and obsolescent.

For research, molecular graphics programs combine the functions of both skeletal and space filling models with high accuracy. The structures may be derived from experimental data, e.g., X-ray single crystal structures, or computed using 'molecular mechanics' or molecular orbital calculations. Such programs will soon be widely available to students.

Expensive models and computer graphics systems, however, may often be replaced by pencil and paper for simple molecules, with a little practice!

2.7   Explanations of the shapes of carbon atoms

So far this chapter has concentrated on the *observed* shapes of organic molecules, to emphasise the importance of experimental evidence. Now let us consider ways of *explaining* the observed arrangements of ligands around C atoms using *either* hybridization *or* molecular orbitals (MOs).

We can make a methane molecule from one carbon and four hydrogen atoms if the eight valence electrons can pair off with opposite spins in each pair. Unfortunately the valence shell atomic orbitals (AOs) of carbon, the non-directional *2s* and the three orthogonal *2p* orbitals, do not lead obviously to the observed shapes of organic molecules. One way to get round this difficulty is to use hybridization to generate directed two-centre two electron bonds, the other is to build multicentre MOs from the AOs.

Hybridization

In the Lewis model of covalent bonding a single bond was equated with a pair of electrons *shared* between two atoms with overlapping atomic orbitals.

Linear combinations of the four valence shell AOs of carbon, one *2s* and three *2p*, subject to limitations on the coefficients in the combinations, are also valid AOs, called *hybridized* orbitals. *Hybridization* is a purely mathematical operation which does not change the distribution of electrons around the atom. It gives a useful picture of directed electron pair bonds. Individual hybrid orbitals are axially symmetrical, with angles between the axes limited to the range 90° to 180° as the contribution of *2p* orbitals decreases from 100 to 50%. There are an infinite number of sets of carbon hybrid orbitals that can be made but three are particularly important.

> If the hybridized orbitals ψ_i (i = 1 to 4) are represented as
> $$\psi_i = c_{i1}s + c_{i2}p_x + c_{i3}p_y + c_{i4}p_x$$
> then $\Sigma c_{ij}^2 = 1$ (j = 1 to 4) for each ψ_i and $\Sigma c_{ij}^2 = 1$ (i = 1, 2, 3, 4) for the AOs.

The *2s* and three *2p* AOs of carbon can give *four equivalent*, axially symmetrical orbitals, called sp^3 ($\equiv s^{0.25}p^{0.75}$) hybrid orbitals, directed from the carbon nucleus symmetrically to the four apices of a regular tetrahedron with C at its centre (Fig. 2.15). The angle between any two sp^3 hybrid orbitals is $\cos^{-1}(-1/3)$, ~109.47°. The four sp^3 orbitals can form *four directional electron pair bonds* with, e.g., four hydrogen atoms to give methane.

> In the general case the coefficients of the AOs contributing to a hybrid orbital are given as superscript decimal fractions following the letter, e.g., *s* or *p*, for the AOs, as in $s^{0.20}p^{0.80}$. For the three special cases given here the whole number *ratios* of coefficients are used.

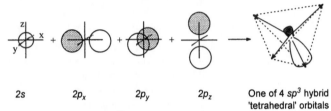

2s	$2p_x$	$2p_y$	$2p_z$	One of 4 sp^3 hybrid 'tetrahedral' orbitals

Fig. 2.15. sp^3 hybridization of carbon orbitals

There are two other special cases of hybridization for carbon orbitals, sp^2 and sp (Fig. 2.16). If one combines the *2s* orbital with *two 2p* orbitals (conventionally p_x and p_y) the axes of the resulting trigonal or sp^2 hybrid orbitals are coplanar, with interorbital angles of 120°. A collinear hybrid pair of digonal or sp orbitals are formed from the *2s* and one *2p* orbital.

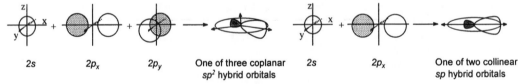

2s	$2p_x$	$2p_y$	One of three coplanar sp^2 hybrid orbitals	*2s*	$2p_x$	One of two collinear sp hybrid orbitals

Fig. 2.16. sp^2 and sp hybridization of carbon orbitals

If two carbon atoms are joined by a σ–bond using two sp^2 orbitals then the $2p_z$ orbitals, each with one electron, can overlap *sideways* to form a π-bond (the σ,π terminology is used in molecular orbital theory). In this way a C–C double bond is formed from one σ- and one π-bond. π overlap of the two $2p_z$ orbitals is best when the axes are parallel and zero when they are orthogonal. Rotating an sp^2 atom with respect to an sp^2 neighbour about a σ-bond effectively breaks a π-bond and the barrier to such rotation is large (260 kJ mol^{-1} in ethene).

When two sp hybrid orbitals on different carbon atoms are used to form a σ-bond the two pairs of *2p* orbitals overlap sideways to form two π-bonds with a *cylindrically symmetrical* electron distribution, as in ethyne.

> Many 4-membered rings are non-planar and must therefore also have bond angles less than 90°

The above picture of C–C single and double bonds is fine for many molecules but small rings (3- and 4-membered) and resonance stabilized systems (see next section) provide problems. We cannot get two carbon orbitals with an interorbital angle $<90°$. A three-membered ring therefore *must have bent C–C bonds* if we wish to describe its bonding in terms of one pair of electrons for each chemical bond. If we allow the amounts of $2s$ and $2p$ orbitals to be varied to maximize the bonding energy we find that bonding in cyclopropane can be described using two $s^{0.20}p^{0.80}$ orbitals for each C–C bond and $s^{0.30}p^{0.70}$ C orbitals for the C–H bonds. At each C atom the interorbital angle of the $s^{0.20}p^{0.80}$ orbitals is $104°$ rather than $109.47°$ (for sp^3) although the *bond* angle is necessarily $60°$. The calculated interorbital angle for the $s^{0.30}p^{0.70}$ orbitals forming the C–H bonds is $116°$, which agrees reasonably well with the *observed* $\angle$ HCH $= 114°$ and which is larger than in, e.g., the $-CH_2-$ in propane ($\angle$ HCH $= 106°$).

Since we cannot avoid 'bent' bonds it should be clear that the contrast between bent bond (see Section 1.2) and σ,π-bond descriptions of the C–C double bond in ethene is more apparent than real. With a suitable choice of hybridization of the carbon atoms in ethene the two models may be made equivalent. Two bent bonds, however, misleadingly suggest a 'hole' between the two atoms where electron density is in fact at its *highest*.

Carbon-carbon triple bonds are *cylindrically* symmetrical about the C–C axis. This is to be expected from the cylindrical symmetry of the electric field produced by the nuclei in a linear molecule such as HC≡CH but it is not immediately apparent from regarding a triple bond as *either* three bent bonds *or* a σ-bond and two π-bonds.

The problem posed by resonance stabilized molecules is dealt with in the following sub-section.

Molecular orbitals

The Lewis model for the formation of an electron pair bond involves a pair of electrons in an orbital formed by the overlap of a pair of atomic orbitals. When a single Lewis structure is not an adequate model for a molecule, e.g., benzene, which does *not* have alternating single and double C–C bonds and is more stable than expected for the Lewis structure, we invoke 'resonance' stabilization: the real molecule is a 'hybrid' of two or more Lewis structures. Molecular orbitals descriptions of conjugated systems give more insight.

Atomic orbitals (AOs) are characterized by wave functions that have *signs*. These signs are irrelevant for orbitals on separated atoms because the electron density is proportional to the *square* of the wave function. The relative signs are important, however, when AOs on different atoms add together to generate *bonding* (BOs) and *antibonding* (ABOs) *molecular orbitals* (MOs: Fig. 2.17; like signs give a BO with increased electron density between the nuclei). Starting with one electron in each AO, with opposed spins, leads to a pair of electrons in the bonding MO. The antibonding MO differentiates this from the Lewis picture of bonding.

N.B. *n* AOs combine to form *n* MOs.

The different signs of lobes of AOs are indicated as hatched or plain in the diagrams. It is irrelevant whether hatched or plain is positive or negative.

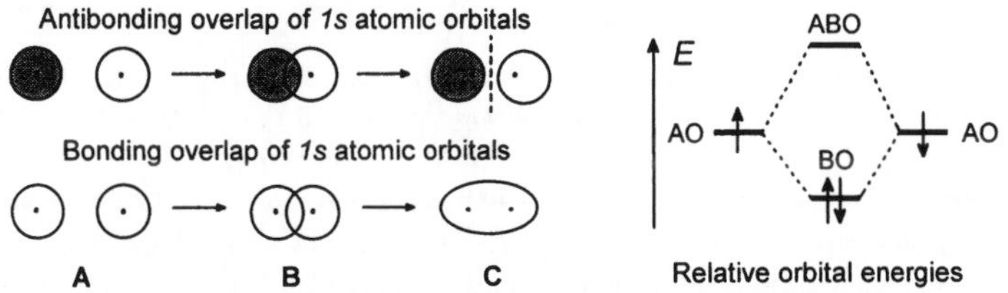

Fig. 2.17. **A**: Atoms with *1s* atomic orbitals (AOs); **B**: Overlapping atomic orbitals; **C**: Bonding (BO) and antibonding (ABO, with a node) MOs.

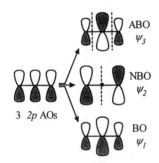

3 *2p* AOs

π-MOs for allyl (propen-3-yl). Broken lines mark nodes in ψ_2 and ψ_3.

Fig. 2.18.

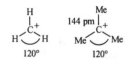

The extent of overlap between a pair of spherical *1s* AOs depends only on the size of the AOs and their distance apart. For *2p* orbitals, with lower symmetry, the overlap also depends on the orientation of the axes of the AOs relative to the positions of the nuclei. I will only consider the sideways or π overlap here. Two *2p* orbitals overlapping sideways give the π-component of a double bond. More interesting is the overlap of three or more *2p* AOs.

MOs may be employed for the whole or *part* of a molecule. We will take the three $2p_z$ atomic orbitals in an allyl (prop-2-enyl) system as an example of the latter (Fig. 2.18), ignoring interactions with the planar 'σ-framework' formed from sp^2 C and *1s* H orbitals. At a simple level the three $2p_z$ atomic orbitals combine to form three MOs, a low energy bonding orbital (BO), ψ_1, a high energy antibonding orbital, ψ_3 (ABO), and a *non-bonding MO*, ψ_2 (NBO), with an energy equal to that of the separate atomic $2p_z$ orbitals. The energy of a pair of electrons in ψ_1 is lower than in the π-orbital of ethene and this energy difference is a measure of the resonance effect in the allyl cation ($CH_2=CH-CH_2^+$). Additional electrons (one or two) will go into ψ_2 (the *non-bonding MO*) without adding to the stability of the system, relative to electrons in atomic $2p_z$ orbitals, i.e., the resonance energy is the same for $CH_2=CH-CH_2^+$, $CH_2=CH-CH_2^{\cdot}$, and $CH_2=CH-CH_2^-$, at this simple level.

The MO description of an allyl system leads very naturally to a symmetrical system with two equal CC bond lengths and there is no need for 'resonance' involving two unsymmetrical structures each with one single and one double CC bond. The allyl system is stereochemically important in two ways. The π-overlap, and therefore the stability of ψ_1, is greatest when the axes of the three *2p* orbitals are parallel. The shapes of ψ_1 and ψ_2 help in explaining the stereo-specificity and -selectivity of [3,3]sigmatropic rearrangements (pages 75–76).

Derivatives of allene (propa-1,2-diene). The central C atom in allene uses two *orthogonal 2p* orbitals to form the two π-bonds, one with each of its carbon neighbours and therefore the two CH_2 groups in allene are orthogonal.

Each double bond resists internal rotation but there is *no cis-trans* isomerism in allenes. Enantiomerism (see Section 4.5), however, is possible with suitable substituents (see also page 2).

The two double bonds in allene are called 'cumulated' and we can extend this to three or more successive double bonds in compounds collectively called *cumulenes* (see **Problems** at the end of the Chapter).

2.8 Bond angles at C atoms in short lived intermediates

Transient intermediates such as carbocations are often difficult or impossible to observe but have a profound effect on the stereochemical outcome of reactions. Our knowledge comes mainly from *ab initio* molecular orbital calculations and from spectroscopy.

Alkyl carbocations (carbon cations)

In the simplest carbocations C^+ is planar (*trigonal* hybridization), as expected for maximum bonding energy, with an *empty 2p orbital*, e.g., as in CH_3^+ and Me_3C^+ (Fig. 2.19). One consequence of this is *racemization* (or *epimerization*) in many S_N1 reactions of chiral alkyl halides (pages 68, 69). For many alkyl cations there are two or more very different structures of similar energy (Fig. 2.20) and such cations are prone to rapid rearrangements through low energy pathways. The structures are 'classical' if they can be described in terms of

Fig. 2.19. Structures of methyl (MO calculations) and t-butyl cations (X-ray study of salt with $Sb_2F_{11}^-$).

two-centred bonds, as in **2.1** or **2.3**, or 'non-classical' if not, e.g., **2.2** or **2.4**.

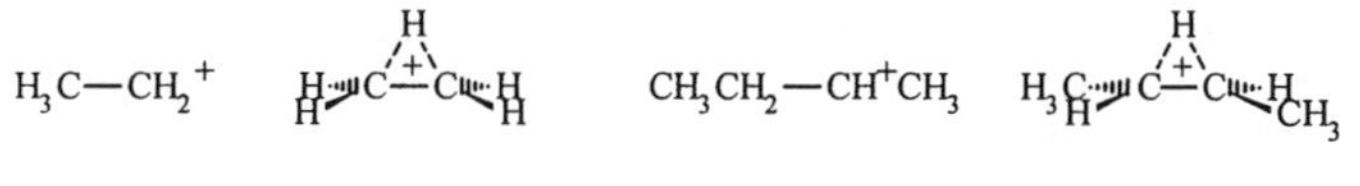

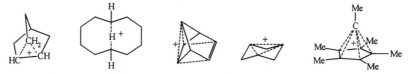

E:	~30	0	10	0
	2.1	**2.2**	**2.3**	**2.4**
	Ethyl cation		2-butyl cation	

Fig. 2.20. Relative gas phase energies E/kJ mol^{-1}(MO calculations)

Although many simple alkyl cations have non-classical structures the longest known and most studied examples occur in cyclic ions (Fig. 2.21).

Fig. 2.21. Examples of cyclic non-classical carbocations: note the 5- and 6-coordinate C atoms in several examples

The vinyl (ethenyl) cation $C_2H_3^+$ is bridged but alkyl substituents change the shape to a *linear* C atom with an empty *2p* orbital (MO calculations; Fig. 2.22). Such ions are relatively uncommon as intermediates.

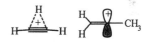

Fig. 2.22. Vinyl cations

Carbon radicals

In an alkyl radical C· is planar or nearly so (the unpaired or 'odd' electron is in *either* a *2p* orbital *or* in a hybrid orbital with very little *s* character), as in $H_3C·$, and *racemization* or *epimerization* accompanies radical formation at a chiral carbon atom.

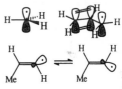

Alkenyl ('vinyl') radicals have the odd electron in an approximately sp^2 orbital but readily isomerize (Fig. 2.23).

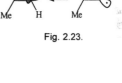

Fig. 2.23.

Carbanions (carbon anions)

Simple alkyl carbon anions, e.g., CH_3^-, are expected to be pyramidal like ammonia but are unknown except for solid alkali metal alkyls from sodium downwards in Group 1 in the Periodic Table. Alkenyl ('vinyl') anions have sp^2 hybridized carbon atoms. Other simple carbon anions such as $^-C≡C^-$ and $^-C≡N$ have the negative charge associated with an sp hybridized σ orbital. Many organometallic compounds are regarded as containing a 'carbanion' so far as organic synthesis is concerned but have complex structures, e.g., $(LiBu)_4$. Often 'carbanion' is used for an anion with the negative charge *mainly* on an electronegative atom such as O, as in an enolate anion (**2.5**), where the geometry is that of an *alkene*.

Carbenes

In simple carbenes one carbon atom is neutral with *six* electrons in its outer shell and has *two nonbonding electrons*. The latter can be paired in a single orbital (a 'singlet state') or unpaired in separate orbitals (a 'triplet state'). The two species have quite different shapes (Fig. 2.24) and often give very different stereochemical outcomes in comparable reactions (page 72).

∠HCH = 104°   ∠HCH = 141°

Fig. 2.24. Singlet and triplet CH_2

Conjugated carbon ions and radicals

In an allyl system the *sideways* overlap of the $2p_z$ (π)-orbitals forces the σ-framework to be coplanar. The nonbonding π-orbital (see Section 2.7) can contain 0 (cation), 1 (radical), or 2 electrons (anion) without qualitatively changing the shape of the allyl system.

Carbon in electronically excited molecules

An organic compound is electronically excited when an electron moves from a bonding (often π) or nonbonding orbital to an antibonding orbital by the absorption of UV or visible light. The change in electron distribution alters bond and torsion angles and this can determine the outcome of *photochemical reactions*. An important example is the photochemical excitation of a bonding π-electron in an alkene to an antibonding π^* orbital. The $2p$ orbitals are parallel in the ground state but *orthogonal* in the excited state. When the molecule changes back to a ground electronic state the $2p$ orbitals reform the π-bond giving *either* the original alkene *or* its E or Z isomer (Fig. 2.25).

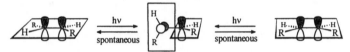

Fig. 2.25. Photochemical isomerization of an alkene

2.9   Bond angles at heteroatoms in organic compounds

The stereochemistry of S in organic molecules is covered by G.H. Whitham, *Organosulfur Chemistry*, Oxford University Press, Oxford, 1995.

Elements in Groups IV and V with four ligands usually imitate sp^3 carbon (Section 2.4). Elements in Groups V and VI with *three ligands* are usually *pyramidal*, with a pair of electrons in a non-bonding orbital (an unshared pair) acting like a fourth ligand on a tetrahedral central atom. The bond angles, however, vary quite widely between $\sim 90°$ and $\sim 120°$ even in 'unstrained' molecules, particularly in nitrogen compounds.

In ammonia the $\angle HNH$ is 104°. If the nitrogen is connected to a π-electron withdrawing group (π-*EWG*) the latter interacts with the *non-bonding* electrons and the angle becomes larger (Fig. 2.26). An extreme case is formamide (methanamide) in which the nitrogen is planar, with a partial double bond between nitrogen and carbon. Electronegative atoms such as halogens (σ-electron withdrawing atoms) pull *bonding* electrons away from the nitrogen, the *non-bonding* pair of electrons is more tightly held in an orbital with greater s character, and the bond angles at nitrogen *decrease* as the p-character of the bonding orbitals increases.

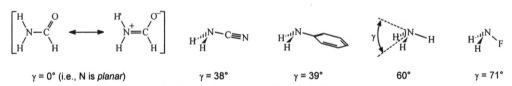

$\gamma = 0°$ (i.e., N is *planar*) $\gamma = 38°$ $\gamma = 39°$ $60°$ $\gamma = 71°$

Fig. 2.26. Deviation from planarity at N (γ is the supplement of the angle between N–X and the bisector of NH_2)

As well as causing wide variations in bond angles at tricoordinate nitrogen these interactions have a dominant effect on the dynamics of *inversion* at nitrogen (page 43) and on the barrier to rotation about the C–N bond (pages 21 and 23).

2.10 Symmetry in finite molecules

What is meant by symmetry? An object is said to possess a symmetry element if a corresponding *symmetry operation* converts it into an object superposable on the original. Reflection in a plane mirror is a familiar example of a symmetry operation. In the Schoenfliess system the symmetry operations possible for finite objects are:

(i) *rotation* through $360° \times m/n$: m is a positive integer < n, about a simple axis of symmetry of order *n* (symmetry element symbol: C_n),

(ii) *reflection* in a plane of symmetry (σ),

(iii) *inversion* through a centre of symmetry (i), and

(iv) a *combination of rotation and reflection* (the order is inconsequential) about a rotation-reflection axis of symmetry of order *n* (S_n). For any S_n axis (*n* = 4 or more) there must also be a $C_{n/2}$ axis.

The plane of symmetry σ is sometimes given the symbol S_1. I prefer σ because there is an infinite number of S_1 axes perpendicular to each σ plane. Similarly there are an infinite number of S_2 axes intersecting each centre *i*.

A *point group* is the sum of symmetry *operations* for a finite object. The distinction between chiral and achiral molecules is important in organic chemistry and chiral and achiral point groups will be considered separately.

Chiral point groups

The symmetry elements in chiral point groups must all be simple rotation axes. The simplest chiral point groups have a single symmetry axis and the point group symbol is the same as that of the rotation axis, apart from the bold face type.

C_1: If a molecule has no symmetry element (except the identity $E \equiv C_1$) it is *asymmetric* and its symmetry point group is C_1. This is common in organic chemistry. It applies to, e.g., all molecules with a *single* chiral carbon atom.

C_n (n >1): Molecules with a single rotation axis C_n are chiral but *not* asymmetric. Point group C_n is rare for n > 2 (see tri-*o*-thymotide, C_3, **5.23**, page 63).

D_n (n >1): In the chiral *dihedral* point groups there is a *principal* axis of order *n* (> 2) and n C_2 axes perpendicular to the principal axis: when *n* = 2 there is no principal axis and the point group is sometimes given the special symbol **V**, rather than D_2.

   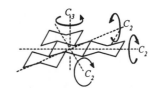

C_1: No symmetry. C_2: A single C_2 axis. D_2 ($\equiv$ V): 3 C_2 axes . D_3: C_3 and 3 C_2 axes.

The most common chiral point groups.

T, **O**, and **I**: These very symmetrical chiral point groups each have *more than one* rotation axis of order n > 2, e.g., **T** has 4 C_3 and 3 C_2 axes. One molecule with symmetry **T** has been synthesised but organic molecules with symmetry **O** and **I** are as yet unknown.

Finite is here used to distinguish these molecules from, ideally, infinitely long polymer molecules.

The symmetry of molecules can be defined exactly, unlike symmetry in everyday objects.
All objects are unchanged by rotation through 360° about *any* axis and therefore have an infinite number of C_1 axes, which is not considered a symmetry element.

Note that the rotation and reflection in S_n constitute a *single* operation.

Point group symbols are usually given in bold type, e.g., T_d and symmetry elements in italic type, e.g., C_3.

Achiral point groups

If the symmetry of a molecule belongs to one of the achiral point groups then that molecule is *identical* with its mirror image. Any point group with one or more of the elements σ, *i*, and S_n is achiral:

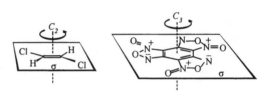

C$_s$: σ plane only. **C$_i$**: Centre of symmetry *i* * **S$_4$**: Rotation-reflection axis
only (this implies a C_2 axis)

Achiral point groups with a *single* element of reflection symmetry

The remaining achiral point groups may be built out of combinations of rotation axes and σ planes. These point groups often have centres of symmetry and S_n axes but I will not refer to them.

C$_{nv}$ and **C$_{nh}$**: These common point groups result when C_n (considered to be vertical) is combined with *either* n vertical σ planes intersecting along the axis *or* a single *horizontal* σ plane perpendicular to the axis.

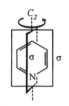
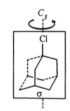
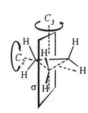

C$_{2v}$: The C_2 axis lies along the intersection of the 2 *vertical* σ planes. **C$_{3v}$**: Only the C_3 axis and one of the 3 σ planes are shown. **C$_{2h}$** and **C$_{3h}$**: the C_2 and C_3 axes are perpendicular to the *horizontal* σ planes. **C$_{\infty v}$**

Examples of **C$_{nv}$** (*n* vertical σ planes intersecting the C_n axis) and **C$_{nh}$** (a *horizontal* σ plane through the C_n axis) point groups

Dihedral groups: D$_{nd}$ and **D$_{nh}$**: These point groups result when σ planes are added to the axes present in **D$_n$**. If the σ planes intersect along the principle axis, they bisect pairs of C_2 axes and are designated as *diagonal* in **D$_{nd}$**. In **D$_{nh}$** there is an *horizontal* σ plane, which must be accompanied by *n*

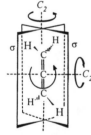
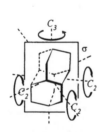
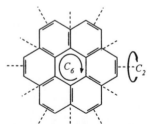

D$_{2d}$: 3 C_2 axes with the 2 σ planes intersecting the principle (vertical) C_2 axis. **D$_{3d}$** : showing only one σ plane through the C_3 axis and bisecting two C_2 axes. **D$_{3h}$** : showing only one vertical σ plane and one C_2 axis through the C_3 axis. The C atoms lie in the horizontal σ plane. **D$_{6h}$** : There are 3 pairs of vertical σ planes (not shown) and of C_2 axes intersecting the C_6 axis. The C atoms lie in the horizontal σ plane.

Two examples of each of the **D$_{nd}$** and **D$_{nh}$** types of achiral dihedral point groups

vertical σ planes that include rather than bisect the *n* horizontal C_2 axes. These are fairly common groups. Ethene (**D₂ₕ**), benzene (**D₆ₕ**), and ethyne (**D∞ₕ**) are other simple examples of the dihedral point groups **Dₙₕ**.

Tᵈ: This very familiar symmetry (of CH_4) results from adding 6 σ_d planes to the rotation axes of **T**.

There are no known organic molecules with the point group Tₕ.

Oₕ: This symmetry is present in the cube and octahedron. These each have 3 C_4, 4 C_3, and 6 C_2 axes and 9 σ planes. Cubane is the only organic molecule with **Oₕ** symmetry, which is common in inorganic octahedral complexes.

Iₕ: This symmetry is present in the dodecahedron and icosahedron. There are 6 C_5, 10 C_3, and 15 C_2 axes and 15 σ planes. Two organic molecules, 'dodecahedrane' ($C_{12}H_{12}$: **2.5**) and 'fullerene' (C_{60}) have this symmetry.

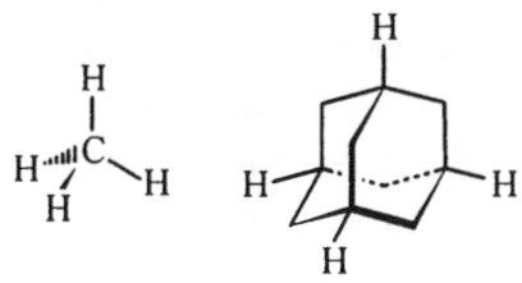

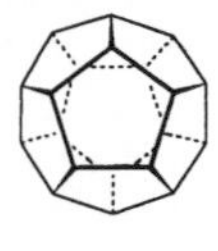

2.5

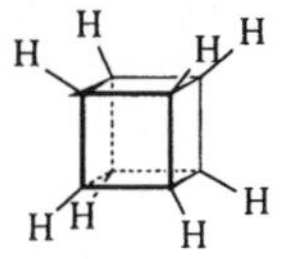

Tᵈ   Oₕ   Iₕ

The examples given above have been rigid molecules with a single conformer (with the exception of **Cᵢ**) to avoid ambiguity. The majority of molecules, however, have two or more conformers, often with different point group symmetries. Such molecules have *dynamic symmetries* that may not correspond to any point group (see pages 45, 46).

2.11 Strain and strain energies: the limits to strain

Whenever bond lengths, or bond or torsion angles, differ significantly from those found in analogous simple acyclic molecules (see pages 5-7), or groups come within sums of van der Waals' radii, a molecule is less stable thermodynamically than might otherwise be expected and is *strained*. It is usually easy in practice to guess whether a given molecule will be substantially strained, although a *quantitative* measure of strain may be difficult to obtain.

Thermodynamic instability does *not* preclude kinetic stability.

Strained acyclic compounds

Few acyclic molecules are severely strained except when there are several large groups such as *t*-butyl (Fig. 2.27). It is usually easy for changes in torsion or bond angles to move large groups apart but bond lengths are not often significantly different (see also molecules **2.6, 2.7,** and **2.26**).

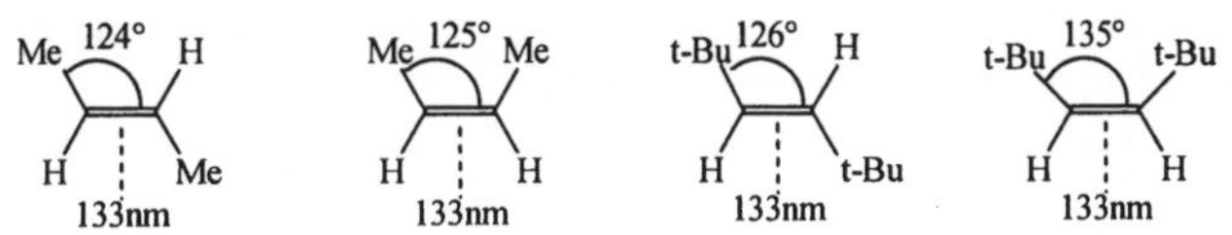

Strain energy difference: 5   39 kJ mol⁻¹

Fig. 2.27. Strain of *Z* and *E* alkenes. Bond lengths and angles are calculated

Ring strain and kinetic stability

Small rings are a common cause of large strain. In cyclopropane the CCC

C–C bond lengths/pm:

151   157

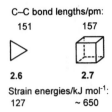

2.6   **2.7**

Strain energies/kJ mol⁻¹:

127   ~ 650

bond angles are ~50° smaller than in propane and the strain energy is about 30% of the strength of, e.g., $H_3C–CH_3$ (350 kJ mol⁻¹). Cyclopropane **2.6** is more reactive in ring opening reactions, because the strain is *reduced in the transition state*, but not in other reactions. Molecules with several small rings may have very large strain energies, e.g. ~650 kJ/mol⁻¹ in cubane **2.7**, which is nevertheless kinetically stable at ambient temperature. Even in these very strained molecules the bond lengths are close to that in ethane (154 pm) (see also Fig. 2.31).

The limits to strain

It is easy to devise molecules that are too strained to exist. *Trans*-cyclo-alkenes are unknown for small rings, even as transient intermediates. *Trans*-cyclohexenes are known as short-lived (~9 μs for **2.10**) intermediates in photo-chemical reactions: the calculated strain energy of *trans*-cyclohexene is ~240 kJ mol⁻¹. Several *trans*-cycloheptenes, e.g., **2.11**, are kinetically stable at very low temperatures. *Trans*-cyclooctene (**2.12**) isomerizes to its *cis* isomer at ~200°C but larger ring *trans*-cycloalkenes differ little from simple acylic alkenes.

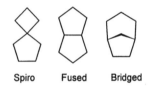

2.8   **2.9**   **2.10**   **2.11**   **2.12**

Fig.2.28. Strained cycloalkenes: **2.8**: all known examples are kinetically stable; **2.9**: *unknown* for n < 6; **2.10** to **2.12**: see text

Spiro   Fused   Bridged

Bicyclic molecules are subdivided into (i) *spiro* compounds, with one atom in common to the two rings, (ii) *fused ring* compounds with two atoms in common, and (iii) *bridged ring* compounds, with three or more atoms in common. Spiro compounds are not highly strained unless one or both the component rings are small. Fused and bridged ring compounds may have much larger strains than might be expected from the component rings.

Strain in fused and bridged bicycloalkanes

Cis-fused and bridged bicycloalkanes are kinetically stable at room temperature for all known ring sizes: (*cis*-)bicyclobutane, the smallest fused system, rearranges to 1,3-butadiene at ~200°C. *Trans*-fused bicycloalkanes must have at least 7 C atoms before *trans*-fused rings are kinetically stable, as in derivatives of **2.13**. Large fused ring systems allow *trans* as well as *cis* double bonds to be common to the two rings in '[*n,m*]betweenanenes', e.g., **2.15**. Racemic [8.8]betweenanene (**2.15**) is quite easily prepared photochemically from the *cis* isomer (**2.14**) but non-racemic chiral analogues have been obtained only by long syntheses.

n

$l+m+3$

$l+m+2$ $l+m+n+2$

1

2

m $l+2$ l

$l+3$ $l+1$

Numbering for bicyclo[*l.m.n*]alkanes ($l \geq m \geq n$), showing the numbers of bridgehead and neighbouring C atoms.

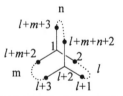

2.13   **2.14**   **2.15**

Bridged bicyclic compounds must have *cis* rings unless at least one chain is long. The smallest kinetically stable *trans* compound appears to be **2.16**. With large bridged systems an additional form of conformational isomerism becomes possible and the *cis* isomers can be either '*out-out*' (as in Fig. 2.29), '*in-in*' (e.g., **2.17**), or *in-out* at the bridgeheads. The bicyclic amine **2.17** forms an *in-in* monocation **2.18** with a very short N···H···N distance as well as an *out-out* dication and an *in-out* monocation with the $^{+}$N–H outside the rings.

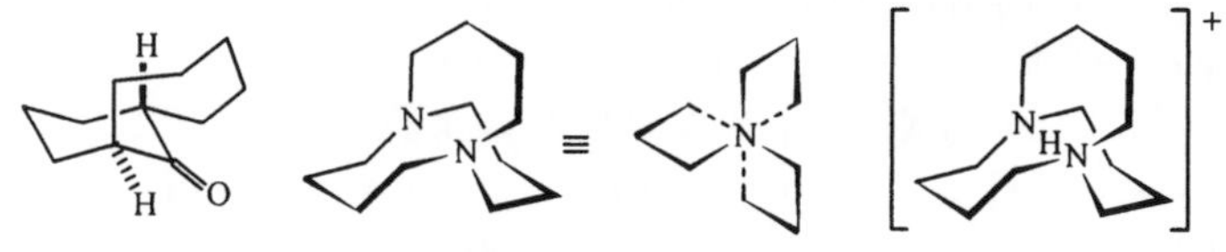

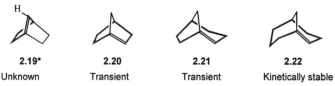

| **2.16** | **2.17** (symmetry **D₃**) | **2.18** |

Fig. 2.29.

An historically interesting example of large strains concerns ⋅ *bridged* bicycloalkenes with a bridgehead double bond. Originally it was thought that such molecules could not exist (Bredt's Rule, 1924). Although there is as yet no evidence for **2.19** many such alkenes can be detected (by trapping with a suitable diene in a Diels Alder reaction: **2.20** and **2.21**) or isolated in suitable cases, e.g., **2.22**. A modern version of Bredt's Rule is that the stability of a given bridgehead bicycloalkene is comparable with that of the most closely related *trans*-cycloalkene (see Fig. 2.30).

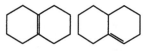

N.B. There is no special difficulty with bridgehead C=C in *fused* ring systems, as in the above examples (derivatives of bicyclo[4.4.0]decane or 'decalin').

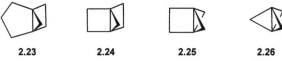

| **2.19*** | **2.20** | **2.21** | **2.22** |
| Unknown | Transient | Transient | Kinetically stable |

Fig. 2.30. Strained bridgehead bicycloalkenes, containing *trans*-cycloalkene sub-units. *The H included in **2.19** emphasises the *trans*-cyclopentene sub-unit.

Cycloalkanes with three or more rings

Tricycloalkanes with three small rings provide a fascinating example of the surprises that chemistry has in store. The tricycloalkane **2.23** (Fig. 2.31) is kinetically stable at room temperature whereas the more highly strained analogues **2.24** and **2.25** are, unsurprisingly, transient. Tricyclopentane **2.26**, however, is readily prepared and stable up to ~200°C!

| **2.23** | **2.24** | **2.25** | **2.26** |

Fig. 2.31. Strained tricycloalkanes. The tricycloalkanes **2.23-2.26** are examples of molecules containing tetracoordinate carbon atoms that are *not* tetrahedral but inverted, with all four ligands in one hemisphere. The C–C bond lenghts/pm in **2.26**, nevertheless, are 152 (C–CH₂) and 160 (C–C).

Tetrahedrane itself (**2.27**; R = H) is unknown and calculations suggest that it has a strain energy of ~540-630 kJ mol^{-1} and is not a kinetically stable system. The tetra-*t*-butyl derivative (**2.27**; R = t-Bu), however, is kinetically fairly stable, rearranging on heating to tetra-*t*-butylcyclobutadiene (**2.28**).

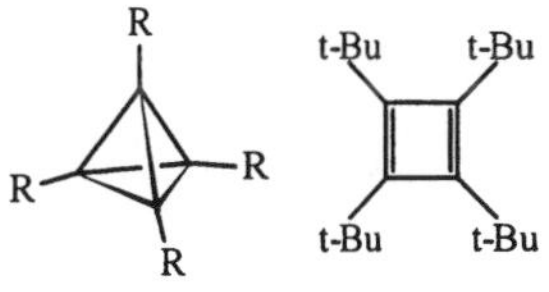

| **2.27** | **2.28** |

Problems

1. Suggest point group symmetries to which the following everyday objects approximate (*n* is an integer): (a) a table fork; (b) a wine bottle; (c) a brick; (d) the seam of a tennis ball; (e) the seam, *including* the stitching, of a baseball; (f) the pair of blades of a Kenwood liquidizer; (g) the teeth of a helical gear wheel with *n* teeth; (h) the spokes of a bicycle wheel with 4*n* spokes; (i) a screw threaded rod; (j) a chopstick; (k) a soccer ball; (l) the dimples on a golf ball. Look out for other everyday symmetrical objects.

2. Assign point group symmetries to the following molecules:

(a) Twistan-2-one; (b) (i) 1-chloro-, (ii) 1,2-dichloro-, (iii) 1,3-dichloro-, (iv) 1,4-dichloro-cubane; (c) the conformers of 1,2-dichloroethane; (d) 4,5-dimethyl-9,10-dihydrophenathrene; (e) 4,5,9,10-tetrahydropyrene; (f) (6)helicene (see Section 4.5); (g) 1,4-dichlorobuta-1,2,3-triene; (h) 1,5-di-chloropenta-1,2,3,4-tetraene.

3. Can you cut a cube into the following numbers of *isometric* parts:

(a) 2 *identical achiral* halves; (b) (i) 2, (ii) 3, (iii) 4, (iv) 5, or (v) 6 *identical chiral* parts; (c) 2 *enantiomorphic* (see page 54) halves?

N.B. *A Guide to IUPAC Nomenclature of Organic Compounds Recommendations 1993*, p. 166, numbers cubane wrongly.

Further reading

J.D. Donaldson and S.D. Ross, *Symmetry and Stereochemistry*, Wiley, New York, 1992.

J.D. Dunitz, *X-ray Analysis and the Structure of Organic Molecules*, Cornell U.P., Ithaca, NY, 1979.

M. Farina and C. Morandi, "High Symmetry Chiral Molecules", *Tetrahedron*, 1974, **30**, 1819.

E. Heilbronner and J.D. Dunitz, *Reflections on Symmetry*, VCH, New York, 1993.

Techniques of Organic Chemistry (ed. A. Weissberger and B.W. Rossiter), Vol. 1, contains accounts of electron diffraction, microwave spectroscopy, neutron scattering.

For discussions of strain energies and strained compounds, see:

A. Greenberg and J.F. Liebman, *Strained Organic Molecules*, Academic, New York, 1978.

W. Luef and R. Keese, *Top. Stereochem.*, 1991, **20**, 231 (strained alkenes)

3   The changing shapes of organic molecules

3.1   Conformations: internal rotation about single bonds

In Chapter 2 we looked at molecules as, in the main, static objects because it is generally difficult to stretch bonds and alter bond angles. In this chapter we consider how molecules can change shape, with emphasis on changing *torsion angles*. There are usually two or more energy minima separated by relatively low barriers for rotation about a single bond. Torsion angles in molecules are therefore inextricably bound up with energy differences and rates of interconversion of the different shapes of molecules.

The different shapes of molecules that result from changes in torsion angles are *conformations* and conformational analysis is the study of the resulting physical and chemical effects. Ethane is a very simple example of conformations (see also Section 1.5). Ethane has three-fold symmetry about the C–C bond. There is a significant energy barrier, ~13 kJ mol^{-1} (Fig 3.1) to the internal rotation of one methyl group relative to the other. This barrier is considerably larger than RT mol^{-1} (~2.5 kJ mol^{-1}) at room temperature and only about 1% of ethane molecules at any instant have enough internal rotational energy to cross this barrier. In comparison with ordinary timescales, however, the average time between crossings is very small, ~10^{-10}s. The barriers to internal rotation about single bonds vary from ~0 to >150 kJ mol^{-1} (see page 44) and the timescales for rotation from ~10^{-12}s to many years, at room temperature.

The lowest potential energy for internal rotation in ethane corresponds to the *staggered* ($\phi = \pm 60°$, 180°) and the highest to the *eclipsed* conformations ($\phi = 0, \pm 120°$) (see also Fig. 2.11, page 8). No ethane molecule is stationary at the point of lowest potential energy and the *zero point energy* for the torsional vibration is ~1.7 kJ mol^{-1}, i.e., ~14% of the barrier (Fig. 3.1). In molecules with torsional energies *above* the barrier the methyl groups rotate relative to each other. For most purposes in organic chemistry barrier heights are adequate and we do not need to know torsional energy levels. Fig. 3.2 gives examples of other molecules with three-fold barriers.

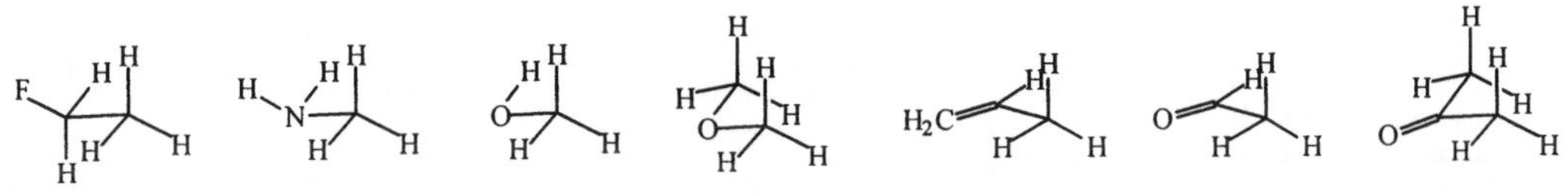

I will not use 'conformation' to include differences in bond lengths and angles, and rotation about *full* double bonds. I will, however, include inversions at pyramidal atoms and rotations about *partial* double bonds as conformational changes.

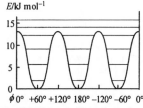

Fig. 3.1. The potential energy of ethane as a function of the torsion angle ϕ for H–C–C–H. The horizontal lines mark the energy levels for torsional vibration states.

Fluoroethane	Methylamine	Methanol	Dimethyl ether	Propene	Ethanal	Propanone
14	8.2	4.5	11.3	8.2	4.9	3.3 kJ mol^{-1}

Fig. 3.2. Barriers to internal rotation (the lowest energy conformation is shown)

Chapter 2 mentioned that X-ray diffraction by single crystals is the most general method for determining the structures of molecules. Unfortunately the

In crystal   In gas
Effect of phase on the conformation of biphenyl.

Electric dipole moments may be measured using the Stark effect, the splitting of rotational energy levels by an electric field.

Vibrational spectroscopy (infrared and Raman) is now mainly of historical interest in studying conformations.

BEWARE VARIATIONS IN USAGE!
A conformer is sometimes called (i) a *topomer* or (ii) a 'preferred conformation', referring to an energy minimum. It has even been used as equivalent to 'conformation'.

3.1

The potential energy minima for internal rotation about the central C–C bond in butane are at or close to $\phi = 180°$ or $\pm60°$, corresponding to staggered conformations, as in ethane.

regularity of crystal structures usually requires that all molecules in the crystal adopt the same shape, which may not correspond to that in a fluid phase. Biphenyl is an example. The methods most suitable for fluid phases depend to an important extent on the size and complexity of the molecules studied.

The most important experimental methods for small molecules in the gas phase have been rotation spectra (in the microwave region and from rotational fine structure in vibrational spectra) and electron diffraction, often in combination. N.m.r. is by far the most important method for solutions (see pages 25–27). It gives less detailed information than electron diffraction or rotation spectroscopy but can be applied to molecules of any size.

Pure rotation spectra are possible only if the molecule has a permanent electric dipole moment (μ): even isotopic substitution can be sufficient, as in H_3CCD_3: $\mu = 0.011D$. A set of rotational energy levels for a molecule gives at most three different moments of inertia. Given sufficient *isotopically* substituted species, however, each with its own set of moments of inertia, it is often possible to determine all the bond lengths, bond angles, and torsion angles of each conformer with high accuracy, particularly if combined with electron diffraction, as well as (fairly low) internal rotation barriers. The complexity of rotational spectra severely limit the size of molecule that can be studied in full detail. The spectrum of propionaldehyde (propanal), with two conformers (see below), took years to analyse.

In addition to experimental methods structures and energies of molecules may be derived from computational methods, particularly *ab initio* MO (best for quite small molecules) and molecular mechanics (Section 3.14).

3.2 Conformers

The torsional potential for the central C–C bond in butane (**3.1**) and most acyclic molecules is quite complex. The words and symbols (Fig. 3.3) each specify a range of torsion of angles (±30°, with signs to distinguish + and − torsion angles) when exact values are not known or are not relevant.

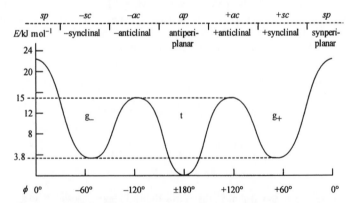

Fig. 3.3. The potential energy of butane (gas phase) as a function of the CCCC torsion angle ϕ illustrating the nomenclature for conformations and conformers. g (for *gauche*) and t (for *s-trans*: see Fig. 2.7 on page 7) are very useful for describing conformers where two or more torsion angles need to be specified qualitatively.

N.B. The selection of atoms or groups defining ϕ is explained in Section 4.7 under **Absolute configuration of chiral axes**.

At room temperature most butane molecules are in one of the three energy wells and it is useful to treat butane as an equilibrium mixture of the three *conformers* (t, g_+, g_-), ignoring the small proportion (<1%) of molecules with torsional energy >15 kJ mol^{-1}. The greater energies of the *synclinal* over the *antiperiplanar* conformation is mainly caused by (1,6)-non-bonding repulsions between hydrogens on the two methyl groups (see Fig. 3.9 on page 26).

Although the barriers to internal rotation are low in, e.g., ethane and butane, the different barriers at $\phi=0°$ (Fig. 3.1 and 3.3) show the important effect of steric hindrance even in butane. In pentane the effects are large enough to eliminate some expected conformers.

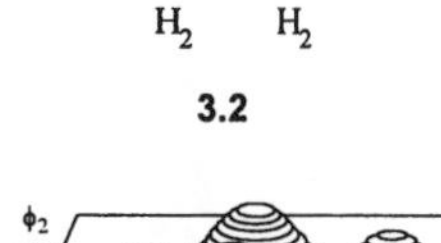

Pentane (**3.2**) requires a new way of representing conformations because there are *two* torsion angles ϕ_1, ϕ_2, which can be used as the axes of a *conformational map*, for C–C–C–C–C sub-units. Frequently potential energies are plotted on the map as contour lines but the result can be confusing. A simpler map results if *populations* rather than energies are plotted, as in Fig. 3.4. The populations decrease regularly from the *tt* conformer ($\phi_1=\phi_2=180°$; with the *lowest* energy) to the *tg* to the g_+g_+ ($=g_-g_-$). The most striking aspect, however, is the *absence* of g_+g_- and g_-g_+ conformers, in which the methyl groups would be very close, and the appearance of *four* very small peaks (see Fig. 3.4) at, e.g., $\sim+90°/-60°$.

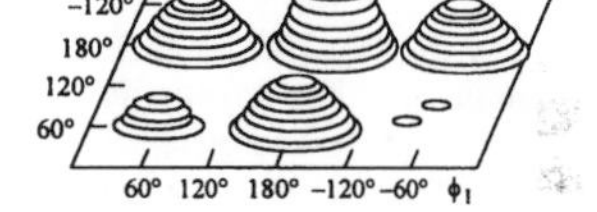

Examples of large barriers to internal rotation will be given in Section 4.5.

3.2

Fig. 3.4.

Non-steric effects in conformational equilibria

Intramolecular hydrogen bonding, e.g., in **3.3**, and electrostatic interactions, e.g., in the neurotransmitter acetylcholine **3.4**, can dominate steric effects in conformational equilibria. The most important examples of such effects controlling the conformations of molecules are found in interactions within and between biopolymer molecules such as proteins, nucleic acids and polysaccharides, and are outside the scope of this book.

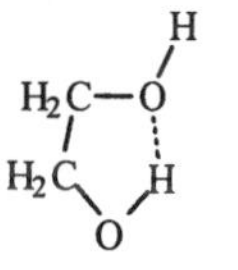

3.3 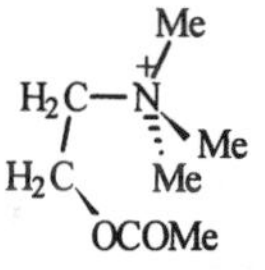   **3.4**

3.3   Molecules with partial double bonds

Torsion barriers can be raised by *resonance* between single and double bond structures, notably in amides and thioamides (Fig. 3.5). The *partial double bonds* increase the torsional barriers and it may become practicable to separate the conformers at room temperature. The barriers in peptidylprolines, e.g., **3.5**, are so important in protein folding that there is a very widespread enzyme, peptidylprolyl-*cis*,*trans*-isomerase (PPI), to catalyse the rotation.

Internal rotation about double bonds in simple alkenes is so difficult (barriers ~ 250 kJ mol^{-1}) that it is usually not considered a conform- ational change (see page 21).

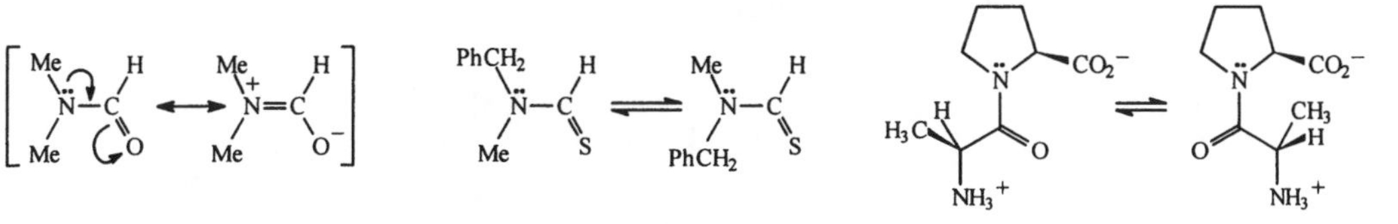

3.5

87 kJ mol^{-1} 105 kJ mol^{-1} 88 kJ mol^{-1}

Fig. 3.5. Barriers to internal rotation about C–N partial double bonds in amides

It is possible for double bonds to be seriously weakened by resonance (so-called push-pull alkenes), with barriers to rotation about the double bonds lowered to values found for some C–C single bonds (Fig. 3.6).

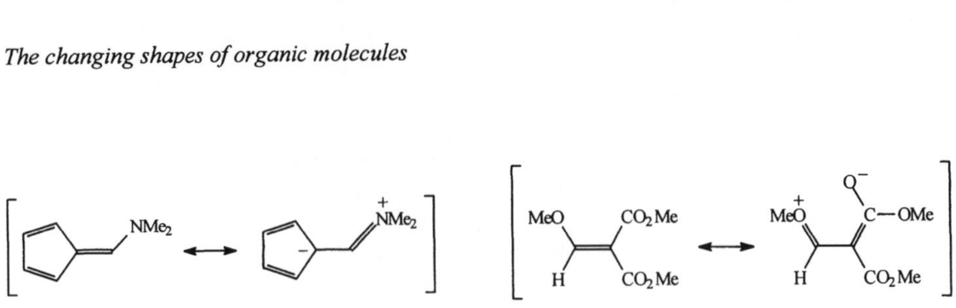

$$\left[\text{structure with NMe}_2 \quad \longleftrightarrow \quad \text{structure with } \overset{+}{\text{NMe}}_2 \right]$$

$$\left[\begin{array}{c} \text{MeO} \quad CO_2Me \\ \\ H \quad\quad CO_2Me \end{array} \quad \longleftrightarrow \quad \begin{array}{c} \overset{+}{\text{MeO}} \quad \overset{O^-}{\underset{|}{C}}-OMe \\ \\ H \quad\quad CO_2Me \end{array} \right]$$

92   116 kJ mol^{-1}

Fig. 3.6. Low barriers to rotation about C–C double bonds in 'push-pull' alkenes

3.4 Introduction to carbocyclic compounds

Most of the remainder of this chapter is devoted to carbocyclic compounds. The omission of most heterocyclic compounds is justified by the availability of good books suitable for students that cover stereoelectronic effects in particular and conformational analysis of heterocyclic compounds in general (see **Further reading**).

Saturated rings with fewer than 12 carbon atoms, except for cyclohexane, have no conformations with bond angles and torsion angles close to those of acyclic analogues. Such rings are significantly *strained*. In 3- and 4-membered rings the strain mainly results from small bond angles, 60° in cyclopropane and 88° in cyclobutane. In cyclopentane there is a compromise between bond angle (average 104°) strain and torsion strain. In 7- to 12-membered rings strain comes mainly from unfavourable torsion angles and repulsive non-bonded interactions between hydrogen atoms. Yet larger rings can have relatively low strain energies. Table 3.1 lists strain energies for 3- to 12-membered cycloalkanes using long chain alkanes as unstrained models.

> Strain energy is the difference between an experimental energy (e.g., $\Delta H°_f$) and the energy of a hypothetical unstrained molecule, choosing the latter is not easy: see Further Reading.

> Normal C–C–C bond angles for chains of methylene groups are about 113° (see Fig. 2.5, page 6).

Ring size, n:	3	4	5	6	7	8	9	10	11	12
Strain (/kJ mol^{-1}):	115	111	26	0	26	40	53	51	47	17
Strain/n (/kJ mol^{-1}):	38.3	27.7	5.2	0.0	3.7	5.0	5.9	5.1	4.2	1.4
Classification:	Small		'Normal'			Medium				Large

Table 3.1. Strain energies in cycloalkanes, from J.D. Cox and G. Pilcher, *Thermochemistry of Organic and Organometallic Compounds*, Academic, New York, 1970, which includes a good account of the assumptions made in deriving strain energies. The classification into small, 'normal', medium, and large (12 or more C atoms) is often used.

The cyclohexane ring is by far the most important saturated ring and, fortunately, the easiest to study.

3.5 Cyclohexane

> Conformations of cyclohexane:
>
>
>
> Chair   Twist   Boat
>
> The twist conformation is chiral: one enantiomer only is shown.

> For axial and equatorial bonds in cyclohexane see page 8.

The conformational equilibrium in cyclohexane

I will use cyclohexane to introduce new stereochemical concepts and terminology and to illustrate applications of n.m.r. in stereochemistry.

The terms 'chair' and 'twist' cyclohexane may refer to *either conformers or conformations*, depending on context. The *boat* is a *conformational transition state* between a pair of enantiomeric twist conformers.

The chair (point group $\mathbf{D_{3d}}$: see Problems, Chapter 2) has the lowest energy of any conformation of cyclohexane and distortion leads to a sharp increase in energy. The six *axial* C–H bonds lean slightly away from the C_3 axis. The six *equatorial* C–H bonds lie *roughly* in a plane perpendicular to the C_3 axis.

The twist conformer is higher in energy and less symmetrical ($\mathbf{D_2}$: it is *chiral*) than the chair. It is flexible and little energy is needed to distort it to the boat conformation ($\mathbf{C_{2v}}$: Fig. 3.7).

Process (c) in Fig. 3.7 is not yet experimentally accessible and the energy difference for the twist and boat conformations has been calculated by molecular mechanics (Section 3.14). Measurement of the activation energies for processes (a) and (b), however, use two techniques for studying conformational changes experimentally that are illustrated below.

If the carbon bond angles were all ideal tetrahedral angles (109.47°) all the bonds would be perfectly staggered. ∠CCC is ~111.4° (compare with propane, 113°, and *g*-butane, 115°).

Relative potential energies/kJ mol^{-1}:

Chair	0
Barrier	~44
Boat	~27
Twist	~22
Barrier	~44
Chair	0

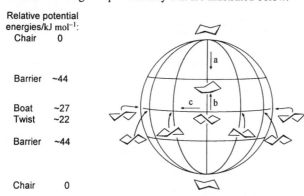

a. Energy rises along meridians from the chairs to the 'tropics' (transition states between chair and twist/boat conformations).

b. Energy rises from twist to the 'tropics'. The energy at the latitude of each tropic is almost constant.

c. The energy along the equator is lowest for the twist and rises to the boat.

Fig. 3.7. A conformational map for cyclohexane on a sphere. Each 'Tropic' corresponds to a 'family' of transition states between the chairs and the twist-boat family of conformations. The chair to twist route along a longitude is via a transition state with four coplanar C atoms shown on the upper tropic,

The twist-boat interconversion along the equator is irrelevant to the chair-chair inversion. Most textbooks use diagrams that wrongly imply that the twist-boat change is a *necessary* part of the chair to chair process.

Cryoscopic trapping of high energy conformers: twist cyclohexane

The high enthalpy of the twist conformers ($\Delta H° = $ ~20–24 kJ mol^{-1}) relative to the chair is opposed by a *favourable* entropy ($\Delta S° = $ +4.9 J mol^{-1} K^{-1}). The concentration of the twist conformer is probably ~0.1% at 300K but rises to ~25% at 800K. If a mixture of cyclohexane and argon at 800K is *rapidly* cooled to ~20K the mixture of chair and twist conformers is frozen in an argon 'matrix'. The rate of the twist to chair change can be measured by infrared spectroscopy at 74K to give the $\Delta G^{\ddagger}$ for process (b) in Fig. 3.7.

The high entropy of the twist conformer relative to the chair is partly due to lower symmetry, partly to greater flexibility. The flexibility is obvious in Dreiding or similar models.

Remember $\Delta G° = \Delta H° - T\Delta S°$.

N.m.r: rates of conformational changes

N.m.r. is uniquely valuable in studying the energetics and dynamics of conformational changes. The energy differences between n.m.r. transitions are very small so that states are distinguishable only if relatively long lived (Heisenberg uncertainty principle). The timescale of n.m.r. experiments (from 1×10^{-6} s to >1s, over a temperature range ~100K to 500K) covers the half-lives of many conformational changes. Each observed chemical shift or coupling constant is an average over a range of molecular shapes and vibrational levels in rapid equilibrium. A separate set of chemical shifts and coupling constants are observed for species interconverting at most slowly. Rates of interconversion of conformers vary with temperature so that chemical shifts and coupling constants may be averages over different ranges of conformations at different temperatures.

An n.m.r. spectrum is characterised by *chemical shifts*, δ (in p.p.m.), *coupling constants*, J (in Hz: interactions between pairs of magnetic nuclei) and *relaxation times*.
Intensities of bands for a given type of nucleus are proportional to the numbers of absorbing nuclei.

Another distinctive feature of n.m.r. is that it can study the rates of processes at equilibrium, as in the chair-chair ring inversion (Fig. 3.7).

Axial and equatorial ^{1}H are in different environments and have different n.m.r. chemical shifts, δ_a and δ_e. Whether separate signals can be observed depends on the relation between $v_0|\delta_a - \delta_e|$ (the spectrometer frequency v_0 depends on the magnetic field and the nucleus), which varies little with temperature, and the rate constant (k), which changes rapidly with temperature, for ^{1}H moving between the two sites (Fig. 3.8).

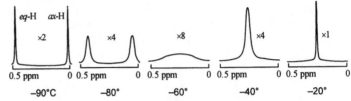

The vertical scales for the spectra are multiplied by the factors shown.

Fig. 3.8. ^{1}H n.m.r. spectra of C_6HD_{11} from –90°C to –20°C, ($v_0|\delta_a - \delta_e|$ = 29Hz).

At –90°C there are two lines, only slightly broadened by exchange ($k \sim 1\,\text{s}^{-1}$) with a chemical shift difference of ~0.48p.p.m. As the rate increases the lines broaden more and move together until they coalesce to a single wide band at ~–60°C ($k \sim 60\,\text{s}^{-1}$). At yet higher temperatures the line narrows until there is little broadening at –20°C ($k \sim 3,000\,\text{s}^{-1}$). The transition state for the chair to chair ring inversion is the half-chair (reached from the chair by process (a) in Fig. 3.7).The ratio of rates of processes (a) and (b) in Fig. 3.7 gives the chair-twist equilibrium constant K (= k_a/k_b).

N.m.r. is also the most important method for studying conformational equilibria in *derivatives* of cyclohexane (next Section).

3.6   Cyclohexanes with a single substituent

See Section 2.5, in particular Fig. 2.13 and the associated text, page 8, about equatorial and axial bonds in cyclohexane.

Most monosubstituted cyclohexanes exist mainly in two *chair* conformers in which the substituent has replaced either an equatorial or an axial hydrogen of cyclohexane. These are referred to as the axial (*ax-*) and equatorial (*eq-*) conformers respectively (Fig. 3.9). Non-bonding *syn*-axial repulsions between an *axial* 1-substituent and *axial* H atoms on C–3 and –5 almost always make the *ax*-conformer the less stable.

The van der Waals radius of H is ~125pm. The H···H distances of 234 pm in *ax-3.6* correspond to significant repulsion.

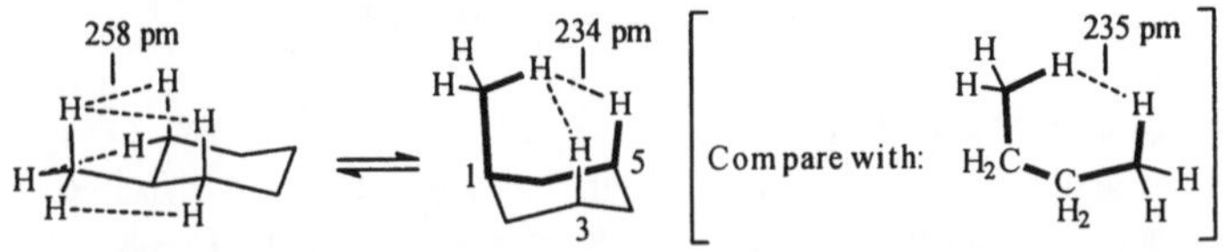

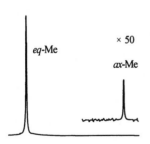

eq-3.6   ax-3.6   a 'gauche butane interaction'

Eq-methylcyclohexane   *Ax*-methylcyclohexane

Fig. 3.9. Chair conformers of methylcyclohexane, with calculated H···H distances.

The best method for measuring the conformational equilibrium in methylcyclohexane (**3.6**) uses ^{13}C n.m.r. at ~173K (the chair to chair change is slow on the n.m.r. timescale). The two conformers give separate signals, identified from empirical correlations (Fig. 3.10 shows the signals for ^{13}C in the methyl groups). K is equal to the ratio of the areas of the two signals (160).

A different n.m.r. approach employs coupling constants ^{3}J (coupling through *three* bonds in H–C–C–H), which are sensitively dependent on the torsion angle, for ^{1}H at high *and* low temperatures. At 165K ^{3}J values for *trans*-H atoms on C–1 and C–2 in the two chair conformers of **3.7** may be

Fig. 3.10. Part of ^{13}C n.m.r. spectrum (proton decoupled) of methylcyclohexane **3.6** at 173K

observed directly. These values may be used in two ways. Firstly, they serve to identify the two conformers qualitatively. The large ^3J for $\phi = \sim180°$ (Fig. 3.11), implies that $-OAc$ is *equatorial* in *eq*-**3.7**. Secondly, they allow the chair-chair equilibrium constant K to be derived from ^3J, a *weighted average* for the two rapidly interconverting conformers, observed at 300K:

Chemical shifts δ usually cannot be used in place of ^3J because δ_e and δ_a change with temperature.

$$K = (^3J_e - {}^3J)/(^3J - {}^3J_a)$$

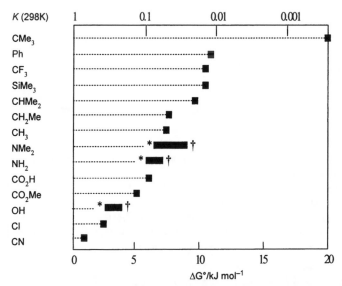

eq-**3.7**   ax-**3.7**

At 165K:
$^3J = 10.4$Hz $^3J = 2.7$Hz
$\phi \sim 180$ $\phi \sim 60$

At 300K:
$^3J = 9.3$Hz: $K = 0.33$

^3J for H–C–C–H versus ϕ
(Karplus 1959)

Fig. 3.11. Conformational equilibrium in acetoxycyclohexane **3.7**. The measurements were made on **3.7** with ^2H on C–3,4,5 with ^2H decoupling in order to get sharp signals.

Fig. 3.12 shows how $\Delta G°_{(e \to a)}$ and $K_{(e \to a)}$ vary with substituents and, in a few instances in which hydrogen bonding is important, with solvent.

K (298K)

	1	0.1	0.01	0.001

CMe$_3$
Ph
CF$_3$
SiMe$_3$
CHMe$_2$
CH$_2$Me
CH$_3$
NMe$_2$
NH$_2$
CO$_2$H
CO$_2$Me
OH
Cl
CN

0   5   10   15   20

$\Delta G°/$kJ mol^{-1}

Fig. 3.12. Chair-chair conformational equilibria in mono-substituted cyclohexanes

(Solvent dependence: * non-polar solvent, † hydroxylic solvent)

$-\Delta G°_{(e \to a)}$ for a monosubstituted cyclohexane is often called an 'A value'. A values are often used as measures of the 'sizes' of groups but can be very misleading for un-symmetrical groups (see Fig. 3.22).

Ax NMe$_2$ has the large Me groups *outside* the ring and the non-bonding electrons on the N directed *into* the ring. The *syn*-3.5-axial H atoms seriously interfere with H bonding with a hydroxylic solvent compared with what is possible for an *eq*-NMe$_2$. The hindrance is less effective for *ax*-NH$_2$ and -OH because H can take up an inside position relatively easily.

The above discussion concentrated on symmetrical substituents, e.g. methyl, but many have low symmetry. With an axial 2-propyl group, as in **3.8**, two of the 'rotamers' (**3.8a** and its enantiomer) are severely strained by a methyl group turned into the ring and only one rotamer (**3.8b**) is significantly populated. The three rotamers of an equatorial 2-propyl group differ little in energy. A detailed study shows that although $\Delta H°$ decreases slightly from methyl to 2-propyl there is a larger increase in $-T\Delta S°$ and $\Delta G°_{(e \to a)}$ increases,

3.8a   **3.8b**

Two rotamers of *ax*-2-propyl-cyclohexane, **3.8**.

3.9

from 7.3 for methyl- to 9.3 kJ mol^{-1} for 2-propyl-cyclohexane.

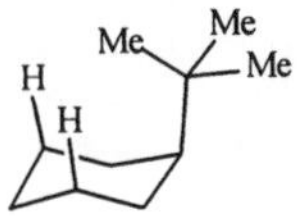

With an axial t-Bu ($\equiv (CH_3)_3C-$) the instability of the *ax* conformer (**3.9**) becomes comparable with twist conformers and for many purposes only the *eq* chair conformer is present. This has important consequences for some disubstituted derivatives.

3.7   Cyclohexanes with two substituents

This Section uses symmetrical substituents in order to concentrate on the stereochemistry of the ring. *Cis* or *trans* is used to characterize the relative configuration of two substituents on the same or on opposite sides of a ring treated as a regular polygon.

What follows describes the chair conformers for cyclohexane with two methyl substituents on different ring atoms, with comments on qualitative changes when the substituents are different. For simplicity the rings are shown with idealized shapes, unlike the realistic distortions shown in Section 3.6.

1,2-Dimethylcyclohexane (3.10)

The chair conformers of the *cis*-isomer (*cis*-**3.10**, Fig. 3.13), are enantiomeric (symmetry point group C_1) but interconvert rapidly by ring inversion ($\Delta G^{\ddagger} \sim 45$ kJ mol^{-1}). *Cis*-**3.10** is therefore an *achiral* compound. With unlike substituents the compound is chiral and K is not unity.

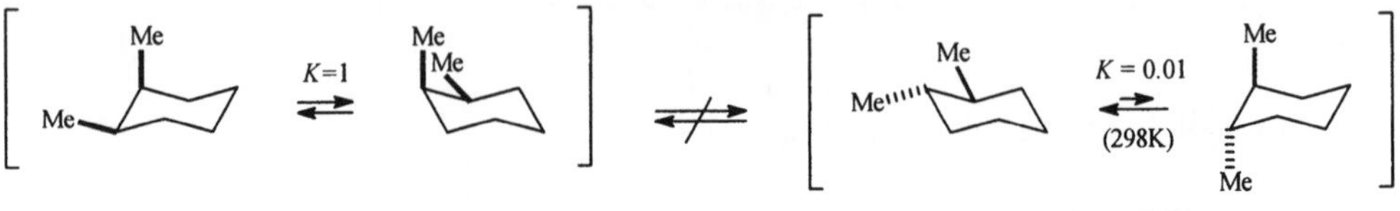

cis-**3.10**   *trans*-**3.10**.

Fig. 3.13. Chair-chair equilibria in *cis*- and *trans*-**3.10**

Trans-**3.10** is chiral but only one enantiomer is shown in Fig. 3.13. The relative energies of the conformers may be estimated *approximately* from the number of *gauche* butane interactions (Fig. 3.9), one in the *e,e*-and four the *a,a*-conformer. *Cis*- and *trans*-**3.10** interconvert only through catalysed reactions, e.g., reversible dehydrogenation over a Pd catalyst at ~300°C.

1,3-Dimethylcyclohexane (3.11)

Cis-**3.11** and both its chair conformers are *achiral* (Fig. 3.14). The *a,a*-conformer is very strained (the Me groups are forced close together, as in the very unfavourable g_+g_- conformation of pentane: see Fig. 3.4, page 23). If there are unlike substituents then the conformers and compound are all chiral. *Trans*-**3.11** is chiral (one enantiomer only is shown in Fig. 3.14). Each enantiomer has only one chair conformer: ring inversion leaves it unchanged! Check this with models.

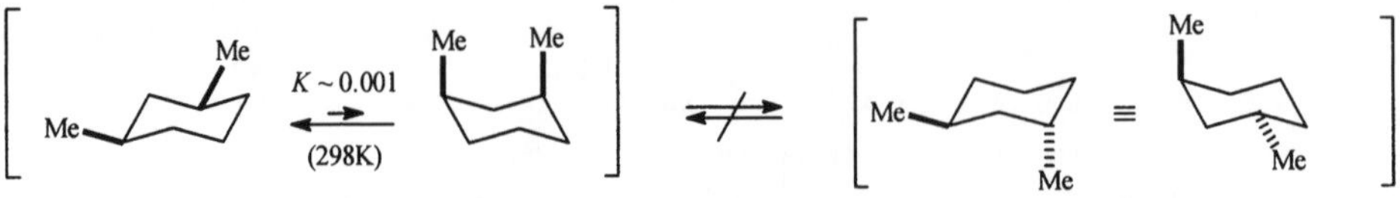

cis-**3.11**   *trans*-**3.11**

Fig. 3.14.

1,4-Dimethylcyclohexane (3.12)

Cis- and *trans*-**3.12** and their chair conformers are all *achiral* (Fig. 3.15). Ring inversion has no effect in *cis*-**3.12**. The strain in *a,a*-*trans*-**3.12** is double that in *ax*-methylcyclohexane (Fig. 3.9).

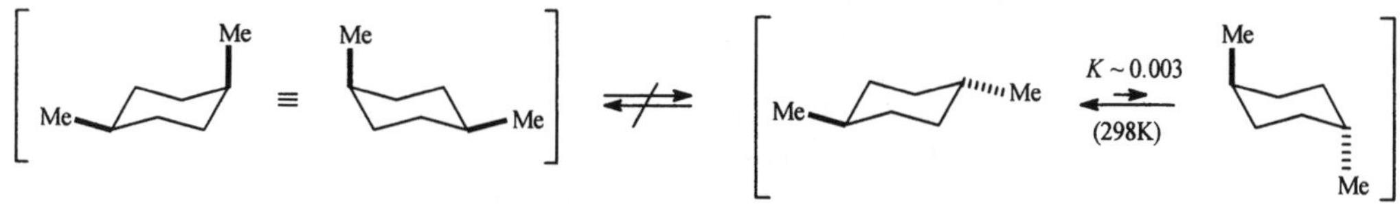

cis-**3.12**

Fig. 3.15.

trans-**3.12**

It might be expected that strain energies would be additive in *cis*-1,4-disubstituted cyclohexanes, an idea supported by molecular mechanics (Section 3.14). This can be put to use in **3.13** to measure $\Delta G°_{(e\to a)}$ for 2-Pr ($\equiv(CH_3)_2CH–$). $\Delta G°_{(e\to a)}$ for Me is well known and $\Delta G°$ for **3.13**, −2.0 kJ mol^{-1}, is easy to measure using ^{13}C n.m.r. at −100°C, and the *algebraic* difference gives $\Delta G°_{(e\to a)} = 9.3$ kJ mol^{-1} for 2-propylcyclohexane.

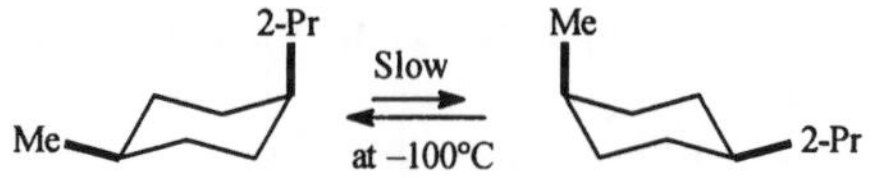

2-Pr

Slow

Me

at −100°C

Me

2-Pr

3.13

Fig. 3.16.

There is no stereoisomerism in 1,1–disubstituted cyclohexanes but one substituent must be *eq* when the other is *ax* and *vice versa*. Conformational energies $\Delta G°_{(e\to a)}$ for the individual groups are *not* additive and may even be *qualitatively* misleading, e.g., for **3.14**. The more stable conformer has *eq*–CH_3, although $\Delta G°_{(e\to a)}$ for $–C_6H_5$ is larger than for Me.

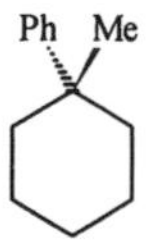

Ph Me

3.14

Me *eq* to *ax*: $\Delta G°= +1.3$ kJ mol^{-1}

Calc.: $+7.3 − 12.0 = −4.7$ kJ mol^{-1}.

t-Butyl as a substituent

When there is one *t*-butyl group and one small or medium sized substituent, e.g., $–OH$ (Fig. 3.17), only conformers with *eq-t*-butyl groups are significantly populated. *Cis*- and *trans*-**3.15**, a readily analysable mixture of stereoisomers unlike the chair conformers of **3.16**, can act as models for the latter in equilibria and, less reliably, in rates of reaction.

OH

t-Bu   OH   Slow   t-Bu   OH   Fast   OH

Cis-**3.15**   *trans*-**3.15**   **3.16**

Slow interconversion via the ketone catalysed by $Al(Oi-Pr)_3$ *Fast* conformational equilibrium

Fig. 3.17.

3.8   Cyclohexanes with double bonds

Cyclohexene and methylenecyclohexane

The two conformers of cyclohexene are the preferred *half-chair* and the *half-boat* (Fig. 3.18). The effect of the double bond is to reduce energy differences compared with analogous changes in cyclohexane and its derivatives. For example, the barrier to ring inversion in cyclohexene is only 22 kJ mol^{-1} (cyclohexane: 44 kJ mol^{-1}). Axial-equatorial energy differences are lower for single substituents, e.g., 4.1 or 3.6 kJ mol^{-1} for 3- or 4-methylcyclohexene (7.3 kJ mol^{-1} for methylcyclohexane). With two substituents, one on an sp^2 C,

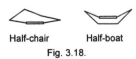

Half-chair   Half-boat

Fig. 3.18.

more striking changes in *ax-eq* equilibria can be observed in derivatives of cyclohexene and methylenecyclohexane, the effects of *allylic* strain.

$A^{(1,2)}$ and $A^{(1,3)}$ (allylic) strain

In allyl systems (the C atoms 1, 2, and 3 in Fig. 3.19) substituents at C-2 and C-3 can greatly change the preferred conformation of the group on C-1.

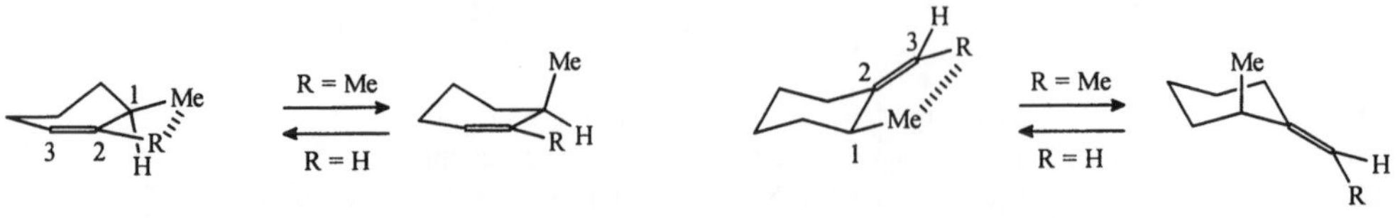

$A^{(1,2)}$ strain $A^{(1,3)}$ strain

Fig. 3.19. $A^{(1,2)}$ and $A^{(1,3)}$ strain, illustrated for methyl groups (repulsive interactions when R = Me are indicated by ⁞⁞⁞⁞)

A-strain can be important even for a partial double bond. When the ketone **3.17** (>90% *cis*, close to the equilibrium composition as ordinarily prepared) condenses with pyrrolidine to form an enamine **3.18**, the latter is >90% *trans* at equilibrium (Fig. 3.20). Kinetically controlled hydrolysis under conditions where enolization is slow then gives mainly *trans*-**3.17**.

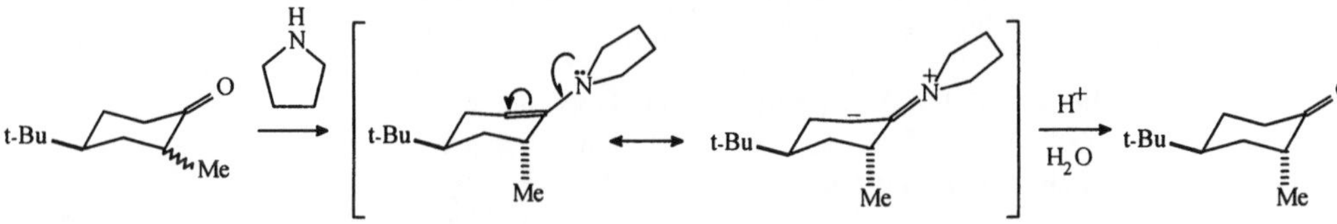

3.17 (~90% cis)   **3.18**   *trans*-**3.17**

Fig. 3.20. An example of $A^{1,2}$ strain resulting from a partial double bond restricting rotation in an enamine

Cyclohexanone

Replacing one CH_2 by C=O has little effect on the shape of the ring but alters many conformational properties, notably the relative stabilities of the chair and twist conformers and of *ax* and *eq* conformers with substituents at C-2.

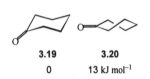

3.19   **3.20**
0   13 kJ mol⁻¹

Fig. 3.21. Relative energies calculated by molecular mechanics

The barrier to rotation about H_2C–C(=O) is lower than for H_2C–CH_2 (see Fig. 3.2). This makes it easier for cyclohexanone than cyclohexane to change from the chair (**3.19**) to twist conformers (e.g., **3.20**: Fig. 3.21) in which unfavourable torsion angles occur for H_2C–C(=O) rather than for H_2C–CH_2.

The C=O affects both polar and non-polar C-2 substituents. The results for 2-alkylcyclohexanones can be explained qualitatively by *decreased* hindrance to axial ethyl (see Fig. 3.22) and 2-propyl groups and *increased* steric hindrance to equatorial 2-propyl (from methyl-O interactions), in the most stable rotamers of the ketones relative to the 2-alkylcyclohexanes.

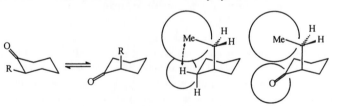

Fig. 3.22. $\Delta G°_{(e→a)}$/kJ mol⁻¹ for 2-R-cyclohexanones (R-cyclohexanes):
R = Me: 7.6 (7.3);   R = Et: 4.5 (7.8);   R = 2-Pr: 1.8 (9.3).

In *trans*-2-chloro-5-methylcyclohexanone **3.21** (Fig. 3.23) the methyl group is hindered when *axial* in *a,a*-**3.21**, while the electrostatic repulsions between the C–Cl and C=O electric dipoles are maximal when the chlorine is *equatorial* in *e,e*-**3.21**. In a non-polar solvent *a,a*-**3.21** is the more stable conformer (lower electrostatic repulsions) but *e,e*-**3.21** is preferred in methanol, which strongly solvates the *larger* resultant electric dipole.

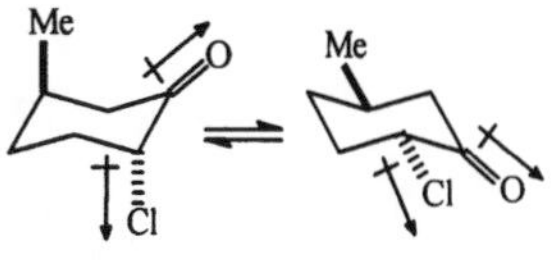

a,a-**3.21**   *e,e*-**3.21**

Fig. 3.23.

3.9 Cyclohexanes in fused ring systems

Cyclohexane rings are common components of *fused* polycyclic systems. Among the most important are the steroids, which include cholesterol (**4.38**, page 48), a major component of nerve tissues notoriously associated with heart disease, vitamin D, bile acids, and sex hormones. An even greater number of di- and tri-terpenes (C_{20} and C_{30} compounds derived, as are steroids, from a common C_5 precursor) include cyclohexane rings.

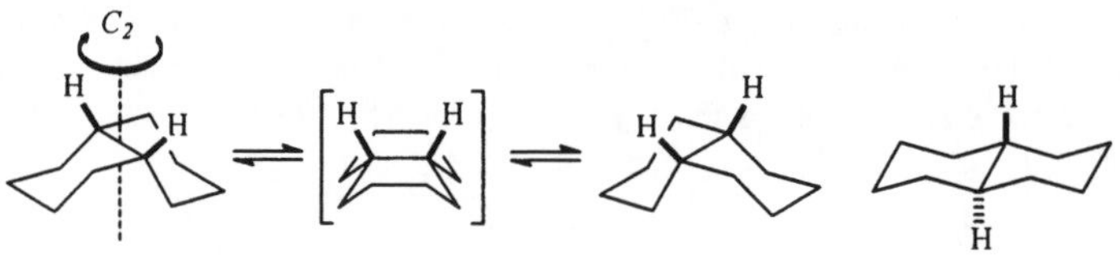

3.22   **3.23**

Fig. 3.24. *Cis*- (**3.22**) and *trans*- (**3.23**) decalin

Cis-decalin (*cis*-bicyclo[4,4,0]decane or *cis*-decahydronaphthalene: **3.22**) has an enantiomeric pair of two-chair conformers, which can interconvert through a two boat intermediate (Fig. 3.24: $\Delta G^{\ddagger} = 51$ kJ mol^{-1}, somewhat higher than in cyclohexane). *Trans*-decalin (**3.23**) exists in a single two chair conformer: neither ring can invert to another chair. With three fused rings another stereochemical distinction must be made (Fig. 3.25). *Cisoid* and *transoid* configurations for the terminal rings specify that the H atoms (or other bridgehead groups) marked with * are *cis* or *trans* on the central ring.

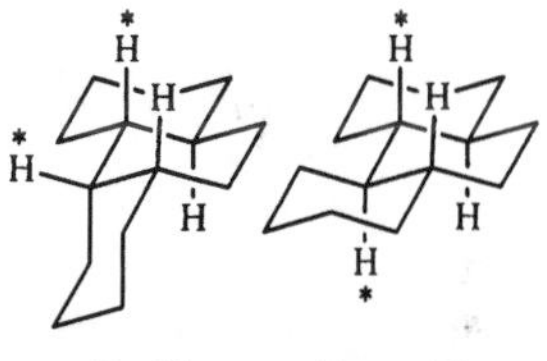

Almost all naturally occurring fused cyclohexane ring systems have at least one *trans* ring fusion and a single all-chair conformer. Many also have 'angular' (bridgehead) methyl groups that (i) often direct attack to the opposite face of the molecules (see Section 6.4 for examples) and (ii) increase *ax-eq* energy differences at the hindered sites (* in Fig. 3.26, the basic ring system of steroids) compared with unhindered sites or with sites in cyclohexane.

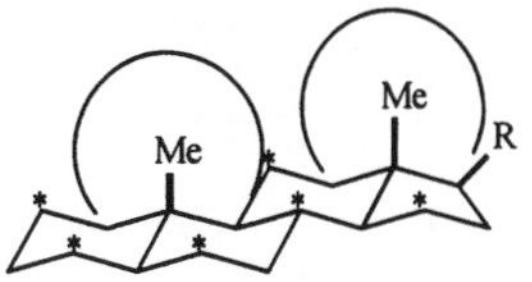

Cisoid   *Transoid*

Fig. 3.25.

Fig. 3.26.

3.10 Cyclic compounds: 3- and 4-membered rings

Three membered rings are necessarily planar so far as the ring atoms are concerned and exocyclic bonds are rigidly oriented. The relationships between the relative configurations and properties of diastereomers are as clear and qualitatively predictable as those of alkenes.

We might expect four membered rings, with an *average* internal bond angle of at most 90°, to be planar in order to minimize angle strain. In cyclobutane **3.24** a planar square conformation has *eclipsed* C–H bonds (torsion angle $\phi = 0°$ for *cis*-hydrogens), which maximizes *torsion* strain. A folded conformation, with a dihedral angle of 28° ($\angle$ CCC = 88°) is lower in potential energy by ~6 kJ mol^{-1}. The CH_2 groups are further tilted, relative to the ring, to lower torsion strain.

3.24

This non-planarity leads to two types of exocyclic bond, similar to the *eq* and *ax* bonds in the chair conformer of cyclohexane. Commonly, however, there are much smaller differences in properties between 'equatorial' and 'axial' conformers in derivatives of cyclobutane. There is a low barrier to ring inversion ($\Delta G^{\ddagger}$ = 6.1 kJ mol^{-1} in cyclobutane) and ~10% of methylcyclobutane molecules will have enough energy to pass freely between the two conformers and therefore are not part of either. Clearly the concept of conformer is less clear in cyclobutane than in butane, let alone cyclohexane.

3.11 Cyclopentane and Cycloheptane

Given that cyclobutane is not planar it comes as no surprise that cyclopentane is also non-planar (average $\angle$CCC = 105°, rather than 108° as in **3.25**, a regular pentagon with symmetry $\mathbf{D_{5h}}$), which is less stable by ~19 kJ mol^{-1}. The 'shape' of a cyclopentane ring is not straightforward. There are symmetrical conformations, the 'envelope' (**3.26**: $\mathbf{C_s}$), the chiral 'half-chair' (**3.27**: $\mathbf{C_2}$) and its enantiomer (Fig. 3.27), and a continuum of $\mathbf{C_1}$ conformations. Although the torsional and angle strains vary as the molecule passes between these conformations the *total* strain energy, ~40 kJ mol^{-1}, is almost constant and considerably less than that of the planar conformation. The C

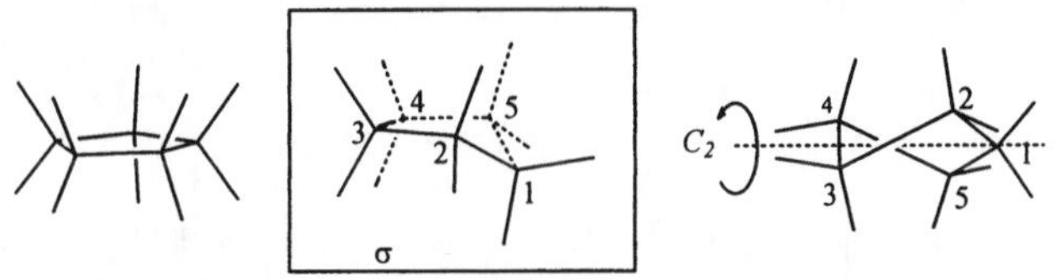

3.25	**3.26**	**3.27**

Fig. 3.27. Three symmetrical conformations of cyclopentane

atoms oscillate *perpendicular* to the mean plane of the ring, with phase differences as a wave passes round the ring. This motion, a *vibration* with almost *no change in potential energy* (contrast with, e.g., any bond stretching vibration) is a 'pseudorotation'.

Any substituent will upset the balance of strains in the cyclopentane ring. A symmetrical substituent will make one of the symmetrical ring conformations the lowest in energy, e.g., $\mathbf{C_s}$ for methylcyclopentane **3.28** but $\mathbf{C_2}$ for cyclopentanone **3.29**. These preferences can be explained by the internal barriers in propane (taken as a model for part of the cyclopentane ring, 14.2 kJ mol^{-1}) and *either* 2-methylpropane (16.3 kJ mol^{-1}) *or* propanone (3.3 kJ mol^{-1}). Placing the methyl group at C–1 in the $\mathbf{C_s}$ conformation *minimizes* the increase in torsion strain. The change from CH$_2$ to C=O *reduces* torsion strain most where eclipsing is greatest in cyclopentane, at C–1 in the $\mathbf{C_2}$ conformation. The pseudorotation of the ring now involves significant energy barriers but these are low compared with ring inversion in cyclohexane.

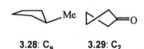

3.28: $\mathbf{C_s}$ **3.29**: $\mathbf{C_2}$

Conformational equilibria in polysubstituted cyclopentanes are hopelessly complicated unless you use a molecular mechanics program and have time to spare.

Cycloheptane has *two* families, 'chair/twist-chair', **3.30**, and 'boat/twist-boat', **3.31**, of pseudorotating conformations (Fig. 3.28) separated by a substantial energy barrier (36 kJ mol^{-1}). Within each family of conformations the *twist* conformations are the most stable, with the chair and boat as conformational transition states. You will not be surprised to find the conformational equilibria in larger rings very complex.

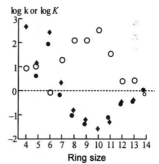

3.30   **3.31**

0	~5	~14	~14
Twist-chair	Chair	Twist-boat	Boat

Fig. 3.28. Relative energies/kJ mol⁻¹ of conformations of cycloheptane

3.12   Medium rings

Conventionally 8- to 11-membered rings are called *medium* and 12- and above *large* rings. Such systems have many conformers that are difficult to study in detail. From 'ball and stick' models the medium and large rings appear to be very flexible and we might expect them to be able to minimize strain quite effectively. Heats of formation, however, show that strain (Table 3.1, page 24) rises to a maximum in the *medium* rings. Calculations show that in these rings one cannot simultaneously have favourable torsion angles, bond angles, and non-bonded distances. Any reaction that reduces the number of ligands on a carbon atom, however, will reduce the non-bonded *transannular* interactions *and* allow less strained bond and torsion angles. Reactions that lower non-bonded repulsions, e.g., S_N1 reactions (open circles: *decrease* in number of ligands at C), are strongly favoured. Reactions that increase non-bonded repulsions, e.g., addition of HCN to, and borohydride reduction of, ketones (filled symbols), are disfavoured. Fig. 3.29 shows these three reactions using rate or equilibrium constants *relative* to acyclic models. From 4- to 7-membered rings there is no obvious pattern. In the medium rings, however, it is clear that the changes in hybridization, and consequent changes in transannular interactions, are dominant.

An exhaustive molecular mechanics search has identified a total of 41 conformers for cyclooctane through cycloundecane within a 25 kJ mol⁻¹ energy 'window' for each ring.

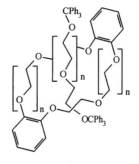

♦: log K for addition of HCN to C=O
●: log k for BH_4^- reduction of C=O
o: log k for S_N1 of tosylates,
all relative to acyclic models
Fig. 3.29.

3.13   Large rings

Rings with 12 or more sp^3 C atoms have too many conformers for any detailed study but very large rings allow new forms of isomerism to arise. A chain can be threaded through a large enough ring and (i) 'capped' with large end groups to form a rotaxane (from the latin *rota*, wheel, and *axis*, axle; Fig. 3.30), (ii) closed to a second ring to form a catenane (from the latin *catena*, chain; Fig. 3.31), or (iii) closed on itself to make a knot. Threading of a polymethylene chain through a polymethylene ring is inefficient, probably because the ring adopts conformations with closely packed parallel segments. Rings based on polyethylene glycol ($HO(CH_2CH_2O)_nH$) are better (**3.32**: a rotaxane). It is far better, however, *either* to have components that positively attract one another and favour a threaded *complex* ready for ring closure (**3.33**) *or* to use a rational synthesis of joined rings (**3.34**) in which late steps break bonds between the two rings (**3.35**: Fig. 3.31).

Catenanes are not just novelties but include, e.g., chained DNA molecules.

Formulae such as **3.32** tend to hide how large the rings are in catenanes and **3.33**-**3.35** (Fig. 3.31) have been drawn out more fully to bring out this point.

3.32
A mixture of chains was used: n is about 8 on average.

Fig. 3.30.

3.14   Molecular mechanics

Chapters 2 and 3 have given many examples of strain and strain energies. It is

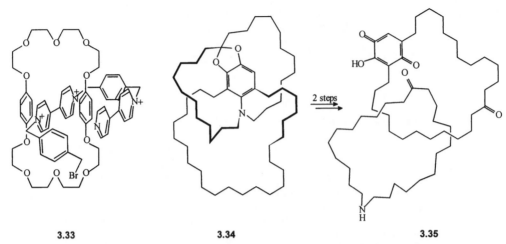

3.33   **3.34**   **3.35**

Fig. 3.31. Formation of catenanes (i) using favourable threading (**3.33**) and (ii) a directed synthesis of *joined* rings (**3.34**) with final hydrolysis (of the ketal) and oxidation (cleavage of the aryl-N bond) completing the synthesis of the catenane (**3.35**)

This account of molecular mechanics applies to relatively small molecules in the gas phase. A more general account of computational methods in chemistry is given by G.H. Grant and W.G. Richards, *Computational Chemistry*, Oxford University Press, Oxford, 1995.

now time to look more quantitatively at strain. What follows is a brief outline of how the potential energy of a molecule varies with changes in bond lengths, bond angles, torsion angles, and non-bonded distances and thence how we may calculate strain energies and other molecular properties using *molecular mechanics*. Its justification is the extent to which it 'works' by reproducing known data and by reliably predicting as yet unmeasured data.

Why don't we simply use quantum mechanics? Unfortunately, high quality *ab initio* molecular orbital (MO) calculations are *very* expensive in computer time and are limited to quite simple molecules. The far quicker 'semi-empirical' MO methods are not very accurate. In molecular mechanics we have a model of how the energies of molecules, *similar in structure to those used to derive the parameters*, vary with change in shape (*without* undergoing chemical reaction) with an accuracy often equal to good experimental data.

This list of terms can be enlarged to include *cross terms* (each with two types of parameter) and solvation, to improve *E*. Cross terms are particularly important for vibration frequencies.

In molecular mechanics the energy E of a molecule relative to a 'strain free' structure is derived from the 'internal coordinates' such as bond lengths and bond angles, and is approximated by a sum of terms, e.g.:

$$E = E(bs) + E(bab) + E(tas) + E(vdW) + E(coul) + etc.$$

where *bs* refers to bond stretching, *bab* to bond angle bending, *tas* to torsion angle strain, *vdW* to van der Waals' (non-bonding) potentials, and *coul* to coulombic interactions. Each of the above terms is itself a sum over, e.g., all bonds for $E(bs)$.

Typically $\Delta E(r)$ is ~15kJ mol^{-1} if $r-r_0$ is 10pm for a C–C single bond.

The simplest treatment of bond stretching uses a quadratic function, $\Delta E(r)$, which is adequate when $|r - r_0|$ is small, for each bond:

$$\Delta E(r) = k_r(r - r_0)^2$$

where r is the bond length at any given instant, r_0 is the minimum energy bond length, and k_r is a constant for the type of bond, e.g., sp^2C–H. In molecular mechanics *it is assumed that such constants can be transferred between molecules*. $E(bs)$ is then the sum of all possible $\Delta E(r)$ terms.

Typically $\Delta E(\theta)$ is ~0.6 kJ mol^{-1} if $|\theta-\theta_0|$ is 4° for ∠CCC: this moves a C atom about 10pm.

A quadratic function is also a first approximation for bond angle bending :

$$\Delta E(\theta) = k_\theta(\theta - \theta_0)^2$$

where θ is a bond angle, θ_0 is the minimum energy bond angle, and k_θ is a constant characteristic of the three atoms defining the bond angle θ. Large bond angle distortions, e.g., in cyclopropane, must be treated differently.

The torsion angle function is naturally a Fourier series. Fortunately such series may often be reduced to quite simple forms, e.g., for sp^3C–sp^3C bonds:

$$\Delta E(\phi) = \tfrac{1}{2}V_3(1 + \cos 3(\phi - \phi_0))$$

where ϕ is the torsion angle defined by a four atom chain X–C–C–Y, ϕ_0 corresponds to the minimum energy, and V_3 is a constant for X–C–C–Y.

Van der Waals interactions have attractive and repulsive terms, e.g.:

$$\Delta E(r) = -A/r^6 + B/r^{12}$$

where r is the distance between two atoms separated by four or more bonds and A and B are constants for this pair of atoms. The attractive term is always $\propto r^{-6}$, which has a theoretical basis, but several functions have been used for the repulsive term, with computational convenience being a major factor.

The term $\Delta E(coul)$ is important for compounds with two or more polar groups. It is may be evaluated from point charges q on atoms i and j, e.g.,

$$\Delta E(coul) = \sum q_i q_j / \varepsilon r_{ij}$$

for atoms separated by four or more bonds, or from bond dipole moments.

The functions for the contributions to E must be provided with appropriate values of the parameters k. Allinger's MM3 parameter set is the best for fairly small molecules, while more specialized sets are available for biopolymers. Each set of parameters is derived from experimental data for a selection of molecules drawn from the range for which the calculations are intended.

A molecular mechanics calculation must begin with an approximate structure for the molecule. This may be derived from measurement of a model (slow!), from a similar molecule, from an X-ray structure database, or from a mouse-drawn diagram on the computer monitor, which the program converts into a 3D model. Computationally the most convenient 'structure' for a molecule with n atoms is a vector of $3n$ Cartesian coordinates $\mathbf{q}$, from which can be derived internal coordinates (r, θ, etc.) needed for the terms $E(bs)$, etc.

The energy E as a function of $\mathbf{q}$ is minimized by a suitable algorithm, almost always using analytical derivatives of E, until a criterion for stopping the process is satisfied and the structure is close to an energy minimum. This may not be the global minimum and finding the latter can be troublesome with complex molecules. For relatively small molecules differences in potential energy calculated in this way, with correcting terms for differences in constitution, may often be useful approximations to differences in heats of formation.

When more than the potential energy E and the shape of the molecule is wanted it is usually necessary to calculate the vibration frequencies, and thence, e.g., entropies and free energies, from the second derivatives of E with respect to the coordinates $\mathbf{q}$ at the potential energy minimum. Unfortunately this procedure assumes that the vibrations are harmonic. It does not handle low energy torsion or inversion barriers, which lead to anharmonic vibrations, or solvation at all well. In *molecular dynamics* one uses numerical methods to calculate the motions of all the particles in the system, which may be a molecule alone or surrounded by solvent molecules, from the forces acting between them. Using averages from such motions followed for long enough gives good estimates of free energies and related properties.

For *ap*-butane $\Delta E(\phi)$ is ~0.15 kJ mol^{-1} if (ϕ–ϕ_0) is 4°: this moves a Me group about 10pm. Torsional constants such as V_3, however, are much more variable than k_r and k_θ.

$\Delta E(coul)$ is troublesome, e.g., it may be necessary to derive the charges from MO calculations. Choosing a value for the dielectric constant ε is also problematic.

It is possible for the minimization to stop at a 'stationary value', such as a saddle point, that is *not* a minimum. If checks show that this has occurred $\mathbf{q}$ must be changed and the minimization repeated until a true minimum is reached.

Problems

1. (a) Explain the number of lines in the ^{1}H decoupled ^{13}C n.m.r. spectra of the compounds listed below at the temperatures/°C given in parentheses.

(i)	*N,N'*-Dimethylformamide	3 (+25)	2 (+150)
(ii)	*cis*-Bicyclo[4.4.0]decane (*cis*-decalin)	5 (−80)	3 (+40)
(iii)	*cis*-1,2-Dimethylcyclohexane	8 (−90)	4 (+25)
(iv)	*trans*-1,3-Dimethylcyclohexane	8 (−90)	5 (+25)
(v)	*cis*-1,4-Dimethylcyclohexane	6 (−90)	3 (+25)

(b) Some pairs of lines in the low temperature spectra in (a) 'coalesce' as the temperature is raised, other lines are unchanged: how many of the *latter* are there for each compound and which C atoms give rise to them?

(c) What are the point group symmetries of (ii) to (v) at low and at high temperatures?

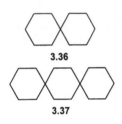

3.36

3.37

Et — Et
Et
3.38

2. (a) Draw spiro[5.5]undecane (**3.36**) with both rings in chair conformations so that both rings look like 'chairs'. (b) What is the symmetry point group of this molecule? (c) What results if you carry out a chair-chair inversion of (i) one ring, (ii) both rings?

3. (a) Draw dispiro[5.2.5.2]hexadecane (**3.37**) with all three rings in chair conformations so that all the rings look like chairs. (b) What is the symmetry point group of this conformer? (c) What results if you carry out a ring inversion on (i) one end ring, (ii) both end rings?

4. How many distinguishable rotamers (see **3.8** and associated text) are there of **3.38**?

Further reading

J.E. Anderson, "Conformational analysis of acyclic and cyclic saturated hydrocarbons", in *The Chemistry of Alkanes and Cycloalkanes*, ed. S. Patai and Z. Rappoport, Wiley, New York, 1992.

G.H. Grant and W.G. Richards, *Computational Chemistry*, Oxford University Press, Oxford, 1995.

A. Greenberg and J.F. Liebman, *Strained Organic Molecules*, Academic, New York, 1978.

J.B. Lambert, "Pyramidal Inversion", *Top. Stereochem.*, 1971, **6**, 1.

Large ring Molecules, ed. J.A. Semlyen, Wiley, New York, 1997.

F.G. Riddell, The Conformational Analysis of Heterocyclic Compounds, Academic Press, London, 1980.

A.J. Kirby, *Stereoelectronic Effects*, Oxford University Press, Oxford, 1996.

N.m.r.; Introductory: P.J. Hore, *Nuclear Magnetic Resonance*, Oxford University Press, Oxford, 1995; advanced: A.E. Derome, *Modern NMR Techniques for Chemistry Research*, Pergamon, Oxford, 1989.

4 Stereoisomerism in molecules and compounds

This chapter begins with a classification of *isomerism* and then goes on to an account of stereoisomerism in molecules and compounds: other forms of stereoisomerism in transition states are covered in Chapter 6.

Classification requires nomenclature. Stereochemistry, unfortunately, has had many changes in nomenclature, which is still developing. I attempt to give an adequate but not pedantic account of current nomenclature.

There is a lack of agreement among chemists about many details of stereochemical nomenclature.

4.1 Classification of isomerism in molecules and compounds

Stereochemical nomenclature: use of prefixes to constitutional names

The *constitution* of a molecule is specified by the sequences of atoms and bonds making up the molecule, e.g., 1-butene and 2-butene are two constitutional isomers of C_4H_8 (Fig. 4.1). In contrast *cis*- and *trans*-but-2-ene (Fig. 4.2) have the same constitution and are *stereoisomeric*. Stereochemical differences are usually specified by *stereodescriptors* (e.g., *cis* and *trans*), prefixes added to names that are based on constitution alone (Section 4.6).

CH_2=$CHCH_2CH_3$	CH_3CH=$CHCH_3$
But-1-ene	But-2-ene

Fig. 4.1. Constitutional isomerism

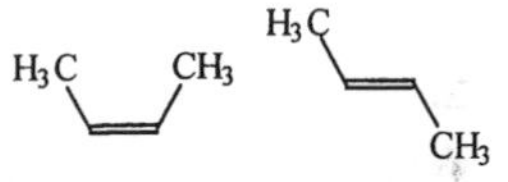

cis-But-2-ene *trans*-But-2-ene

Fig. 4.2. Stereoisomerism

Figure 4.3 shows the possible relationships between *two molecules with the same molecular formula*.

Superposable?	No →	Related as object and mirror image?	No →	Same constitution?	No →	Constitutionally isomeric
Yes ↓		Yes ↓		Yes ↓		
'Homomeric' (rarely used)		Enantiomeric		Diastereomeric		

Fig. 4.3. Relationships between two molecules with the same molecular formula

Diastereo*is*omer, diastereo*is*omeric, etc., are sometimes used in place of diastereomer, diastereomeric, etc.

Two molecules or compounds may be shown to be enantiomeric (or not) experimentally, *independent of any theory of chemical structure* (see Section 1.1). Diastereomeric molecules or compounds differ in *all* properties that depend on structure. It is only when we define the constitutions of molecules in terms of single, double, triple bonds, etc., that we can distinguish between diastereomers and constitutional isomers. For example, *cis*- and *trans*-butenedioic acid, with striking differences in properties, were originally given different names, maleic acid and fumaric acid, implying different constitutions, before van't Hoff (1874) proposed that they were stereoisomers.

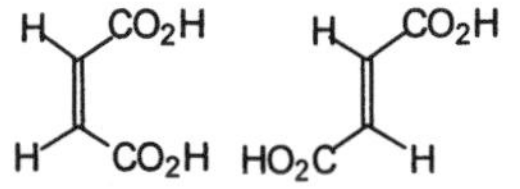

Maleic acid	Fumaric acid
M.p. 139°C	M.p. 286°C
pK_a 1.8, 6.6	pK_a 3.0, 5.0
788g dm^{-3}	7g dm^{-3}

(solubility in H_2O at 25°C)

Always use *structure* to include *stereochemistry*, otherwise use the term *constitutional isomer*.

The relationships between the pairs of terms 'enantiomeric/diastereomeric' and 'enantiomer/diastereomer'

It may be helpful to look at family relationships. A person may be parent, child, cousin to various other persons but cannot have the relationships of parent *and* child *and* cousin with the same person.

Fig. 4.3 serves to define enantiomeric and diastereomeric *relationships* as mutually exclusive. The derived words enantiomer and diastereomer, however, are *not* mutually exclusive. For example, there are three stereoisomers of tartaric acid (Fig. 4.4): (+)-tartaric acid is both the enantiomer of (−)-tartaric acid and a diastereomer of *meso* tartaric acid: which of the two terms is used for (+)-tartaric acid depends on the context.

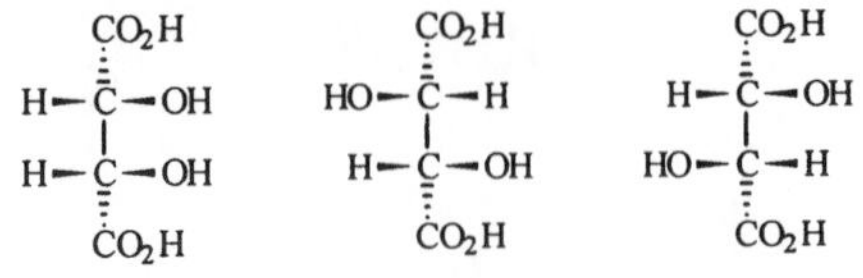

$$
\begin{array}{ccc}
CO_2H & CO_2H & CO_2H \\
H-C-OH & HO-C-H & H-C-OH \\
H-C-OH & H-C-OH & HO-C-H \\
CO_2H & CO_2H & CO_2H \\
\end{array}
$$

Prefix: *meso*   (−)   (+)

Fig. 4.4. Stereoisomers of tartaric acid

Absolute and relative configurations

The relationships in Fig. 4.3 provide a basis for dividing stereodescriptors into two sets. Two diastereomers differ in the distances between atoms and their relationship is not changed if one is reflected in a plane mirror: such differences are *geometric*. Enantiomers do not differ in the distances between constitutionally equivalent atoms (they are *isometric*) but their relationship is changed if one is reflected in a plane mirror: they differ *topographically*. Fig. 4.3 leads to the following correlation between classification of isomerism and the mathematical description of the differences between the classes of isomer:

The use of upper case for *Z* and *E* was established before the general rule specifying lower case letters.

Isomerism:	Constitutional	Diastereomerism	Enantiomerism
Differences:	Topological	Geometrical	Topographical
Descriptors specify:	—	Relative configuration	Absolute configuration
Style of descriptor:	—	Lower case letter (*except Z and E*)	Upper case letter

The classification in Fig. 4.3 for molecules can be applied to *physical (non-bonding) interactions*, e.g., between two molecules, or between a molecule and a chromatographic absorbent, and so on. If we modify constitution to allow partial bonds at the sites of reactions we can also apply the classification of isomerism to *transition states* (see Chapter 6).

4.2 Molecules and compounds

For our purposes a molecule is a microscopic grouping of atoms held together by bonds and stable to small relative displacements of the atoms.

Contrast this with a sample of a compound, a macroscopic assembly of molecules homogeneous to methods of separation, e.g., crystallization, distillation, or chromatography, which are slow compared with many, but not all, of the reversible changes possible for molecules.

A compound may be made up from one type of molecule only (neglecting isotopes), e.g., methane contains only tetrahedral CH_4 molecules. Far more commonly a compound will contain several conformers which are distinguishable using spectroscopic and diffraction techniques with a short timescale («1s) but are not separable, e.g., by distillation or chromatography. There are 'grey' areas, e.g., the barrier to rotation in derivatives of biphenyl (Section 4.5) can vary from ~ 0 to > 180 kJ mol⁻¹ (Fig. 4.13). For the lower barriers rotation is fast but barriers > 90 kJ mol⁻¹ can give stereoisomers separable at room temperature. These grey areas are unavoidable.

4.3   Chirality in compounds

Students often misunderstand the limitations of physical methods for studying molecular structure and this is true of chirality. Most organic molecules are far too complex for detailed study by spectroscopy or electron diffraction but even for simple molecules these methods are limited. For example, an electron diffraction pattern depends only on distances between atoms and gives *geometrical* information. It cannot, however, determine the *topography* of molecules in a sample of a compound, i.e., whether one or both enantiomeric forms of a chiral molecule are present (in equal or unequal amounts). Under suitable conditions, however, X-ray (Section 1.6) and neutron diffraction on single crystals, certain nuclear magnetic resonance experiments, experiments with *polarized* light, and a few specialized techniques applicable to certain crystals, allow us to determine the handedness of chiral molecules or distinguish between chiral and achiral compounds.

A compound composed of chiral molecules of one handedness only (within experimental error) is called *enantiopure* or *homochiral*; the latter is used by most chemists although it has a second quite different meaning. A compound made up of chiral molecules *X*, of one handedness, is said to be enantiomeric with the non-identical 'mirror image' compound made up of chiral molecules enantiomeric with *X*. Such compounds are related as object and non-identical mirror image in the sense that *any chiral arrangement* of a group of molecules in a compound is as *probable* as its mirror image in the enantiomeric compound.

Detection of chirality in compounds: plane and circularly polarized light

The most widely used methods for detecting chirality in compounds use *polarized* light to measure optical activity and circular dichroism. When the sensitivity of these *chiroptical* methods is inadequate, (bio)chemical, n.m.r., and chromatographic methods may be used (Chapter 5).

Light is electromagnetic radiation propagated as waves with orthogonal electric and magnetic vectors. The orientation and magnitude of the electric vector in ordinary light from, e.g., a filament lamp vary randomly in time and space and such light is *non-polarized*.

In *plane* (or *linearly*) *polarized* (*p.p.*) light the electric vector is constant in orientation but varies sinusoidally in amplitude with a frequency ν: such light is *achiral*. In *circularly polarized* (*c.p.*) light the electric vector is constant in magnitude but rotates, with an angular velocity 2πν radians s⁻¹, as it moves forward: such light is *chiral* because the electric vector traces out a helical (chiral) path that may be *either* right *or* left handed.

Optical activity

In a chiral medium the refractive indices for left, n_l, and right, n_r, *c.p.* light are different but it is rarely practical to determine n_l and n_r separately.

In organic chemistry 'chiral compound' always implies chirality in a *fluid phase*: contrast this with chiral *crystals* that become achiral on melting, e.g., quartz, or in solution, e.g., $NaClO_3$.

Most spectroscopic methods have limitations similar to those for electron diffraction.

'Optically active' is not acceptable as an alternative to homochiral. A sample may be optically active simply because there is an unequal mixture of enantiomeric molecules. In other instances optical activity may be undetectable in a sample of a homochiral compound.

Some textbooks say that plane polarized light is *chiral*! Its use, *in conjunction with a chiral polarimeter*, to detect chirality depends on it being achiral.

The magnitude of most measurements of rotation is <50° and correspond to $|n_l - n_r|$ <1.6×10^{-8}, with a sensitivity better than 1×10^{-12}, for the sodium D line.

Optical activity (o.a.) in chiral substances is sometimes referred to as *natural* o.a. to distinguish it from o.a. *induced* in an achiral substance by a magnetic field (Faraday effect).

$$\left[\alpha\right]_{\lambda}^{T} = \frac{\alpha_{\lambda}^{T} \times 100}{l \times c} \quad or \quad \left[\alpha\right]_{\lambda}^{T} = \frac{\alpha_{\lambda}^{T}}{l \times d}$$

A typical polarimeter cell is 10 cm (= 1 dm) long. Historically dm has been used as the unit of length, with concentrations in g/100 cm³.
λ is given in nm *or* as a symbol such as D for the emission line of a sodium lamp at 589nm.

Enantiomeric excess is sometimes called enantiomeric purity but the latter term has been used in more than one sense and is best avoided.

A simple way of detecting chirality in a pure compound, given an adequate sample, is to determine whether or not the compound is *optically active* (*o.a.*), i.e., rotates (or, better, *twists*) the plane of polarization of *p.p.* light. Chirality in a compound may also be detected by *circular dichroism*, i.e., a difference in intensity of *absorption* of left- and right-handed *c.p.* light.

If a pure compound in a fluid phase shows detectable optical activity then that compound must be chiral and non-racemic. The converse is not true. If the sample is very small (e.g., in biochemical work) or the optical activity expected may be inherently small (e.g., when the chirality results from isotopic differences), or the rotation is coincidentally zero at the wavelength of light used (see, e.g., Fig. 4.5), optical activity may be undetectable.

P.p. light is achiral but when it passes through a chiral medium (gas, liquid or solid) the plane of polarization is *twisted* and it is this (chiral) twisting that is detected by a polarimeter: it is the *polarimeter that is the chiral object* that serves to detect the chirality of the path traced out by the *p.p.* light as it passes through the optically active sample.

Optical activity is usually measured using solutions of organic compounds, although pure liquids are also suitable. The optical rotation, $\alpha°$, varies with the path length, l, the wavelength of the light, λ, the temperature, T, and *either* the density d of a pure liquid *or* the solvent and concentration c (g/100 cm³) of a solution, as well as the substance. The *specific rotation* $[\alpha]$ is derived from α, l, and c or d: its dimensions are deg cm² g⁻¹ (which are usually omitted), *not* deg. A specific rotation should be given as, e.g., −14.1, *not* as −14.1°.

$[\alpha]$ would be independent of c for ideal solutions but the concentrations are usually too high for ideal behaviour. It is usual, therefore, to quote c as well as the solvent in parentheses following the value of $[\alpha]$ for solutions.

A polarimeter cannot tell α from any angle $\alpha \pm n \times 180°$ (*n* is an integer). The ambiguity can be resolved by changing *l* but this is rarely necessary. From experience *l* and c are chosen to ensure that α is at most a few degrees.

When making comparisons of optical activity for chemically related chiral compounds it is usual to use the *molar rotation*, $[\Phi] = [\alpha] \times RMM/100$, where RMM is the relative molecular mass (molecular weight).

Chiral liquid crystals have very high rotations (up to 10^6 °/dm!) but must be measured as pure liquids because dilution destroys the order that gives rise to the very large rotations.

Optical purity and enantiomeric excess

Apart from showing qualitatively that a sample of a chiral compound is not racemic, what is the use of $[\alpha]$? One important use is to determine the *enantiomeric* composition of a sample. If the sample is *o.a.* (and free from optically active impurities) it must have an excess of one enantiomer. If the rotation $[\alpha]_{(100\%)}$ is known for that enantiomer then the *optical purity* (*o.p.*) of the sample is:

$$\% \ o.p. = [\alpha] \ (\text{sample}) \times 100 \ / \ [\alpha]_{(100\%)}$$

The enantiomeric excess % *e.e.* is given by:

$$\% \ e.e. = |\text{mole fraction of (+)-form} - \text{mole fraction of (−)-form}| \times 100$$

It is usual to assume that % *e.e.* = % *o.p.* but this is an approximation. If there are strong solute-solute interactions the difference between *o.p.* and *e.e.* can be large (the Horeau effect, first observed in **4.1**).

4.1

If [α] is *not* known for the enantiomerically pure compound or is too small to measure then *e.e.* must be measured by other methods (pages 64, 65).

Optical rotatory dispersion and circular dichroism

A refractive index varies with wavelength (*optical dispersion*). Similarly [α], which is related to $n_l - n_r$, varies with wavelength (*optical rotatory dispersion, ORD*). Chiral compounds exhibit *anomalous rotatory dispersion* (the *Cotton Effect*) and *circular dichroism (CD)* in an ultraviolet or visible absorption band (Fig. 4.5). The absolute configurations of some chiral compounds can be inferred from the *signs* of Cotton Effects and CD using empirical rules.

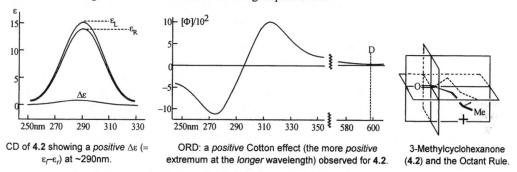

CD of **4.2** showing a *positive* Δε (= $ε_l$–$ε_r$) at ~290nm.

ORD: a *positive* Cotton effect (the more *positive* extremum at the *longer* wavelength) observed for **4.2**.

3-Methylcyclohexanone (**4.2**) and the Octant Rule.

Fig. 4.5. CD and ORD of (+)-3*R*-methylcyclohexanone, **4.2**. N.B. An *extremum* is a minimum *or* maximum

Fig. 4.5 gives the Cotton Effect and CD for the n→π* transition for the C=O group (the 'chromophore') in (+)-3*R*-methylcyclohexanone (**4.2**). Such 'forbidden' transitions were among the first to be studied because the weak absorption lets light through the sample, and the CD Δε is relatively large. In more complex molecules there may be overlapping CD bands and Cotton Effects, as well as vibrational fine structure.

The Octant Rule was originally derived empirically but also has a theoretical basis from *approximate* MO theory.

The Octant Rule uses three surfaces to divide the space around a C=O into 8 regions, with alternating signs across any surface: one positive near-side octant only is marked in Fig. 4.5. A chiral ketone will have at least one atom or group more polarizable than hydrogen that is not matched by another in an octant of opposite sign: the Me in **4.2** is such a group. The CD and Cotton Effect are positive and therefore the Me group is in a positive octant, which is possible only if **4.2** has the *R* configuration. The Octant rule, and analogous rules for other functional groups, is not very reliable and fails, e.g., for **4.3**. CD, however, is valuable in studying chiral polymers.

4.3

4.4   Stereogenic units in molecules

A molecule has a stereogenic unit (e.g., a *chiral* centre (**4.4**), axis (**4.5**), or plane **4.6**, or a double bond (**4.7**), or *some* tri- or tetra-coordinate atoms lying on symmetry planes (**4.8**)) if exchange of an appropriate pair of ligands generates a stereoisomer.

4.4   **4.5**   **4.6**   **4.7**   **4.8**

Fig. 4.6. Molecules with a single type of stereogenic unit

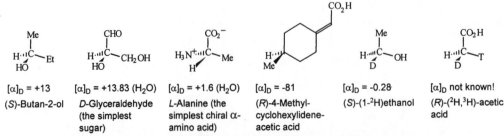

Chiral centres such as the C atom in **4.4**, with four different ligands (see Fig. 4.7) or *pyramidal* atoms with three different ligands (see Fig. 4.11), are the most common chiral unit. The C=C=C unit in **4.5** constitutes a *chiral axis*, about which there is a chiral distribution of ligands. In **4.6** the carboxyl substituted benzene ring is a *chiral plane* (see Section 4.7): exchange of the –CO$_2$H with any one of the H atoms shown explicitly gives the enantiomer. **4.7** and **4.8** are examples of achiral diastereomers.

The molecules in Fig. 4.6 each have a single stereogenic unit. Most molecules showing stereoisomerism are less simple and have two or more regions of the molecule with *independent* stereogenic units: distinguishing such units is *factorization*. The naturally occurring lactone **4.9** has four stereogenic units that can be altered *independently*: C-6 and C-8 are chiral centres; C-10,1,2 defines a chiral axis, and the C-4,5 double bond can give *Z/E* isomerism. The C=C at C-7, however, cannot generate stereoisomers because the five membered ring makes it physically impossible to interchange ligands at one end or the other. In a larger ring (see Section 2.11) a C=C can be a stereogenic unit, like C(4)=C(5) in **4.9**. Similarly small bridged bicyclic molecules must be *cis*-bridged (Section 2.11) and the configuration of bridgehead atoms, e.g., C–1 and C–4 in **4.10**, cannot be changed independently. Thus there are only two (enantiomeric) stereoisomers of **4.10**.

4.9

4.10

4.5 Structures of molecules of chiral organic compounds

A pure compound may be homochiral if (i) the constituent molecules are chiral and (ii) there is no low energy route for the interconversion of enantiomeric molecules. Fortunately, common structural features regularly satisfy both conditions in the great majority of chiral compounds. In other, less common, classes of compound the rates of interconversion of enantiomeric molecules varies greatly with small differences in structure.

The recognition of chiral units in a molecule, subject to certain restrictions, is often the simplest way to determine whether or not a compound will be chiral. This Section deals with compounds with a single chiral element; more complex examples follow in Section 4.6.

Chiral centres: chiral carbon atoms

Chiral is strongly preferred to the obsolete 'asymmetric'.

A tetrahedral atom, or a pyramidal atom with three ligands (an unshared pair of electrons may be regarded as a fourth 'ligand'), is a *chiral centre* if the interchange of any two ligands leads to a new stereoisomeric molecule.

A molecule with a single chiral centre must be chiral *but* molecules with *two or more chiral centres* are *not* all chiral (see Section 4.6).

Me H··ᐟᐟᐟC—Et HO	CHO H··ᐟᐟᐟC—CH$_2$OH HO	CO$_2^-$ H$_3$N$^+$··ᐟᐟᐟC—Me H	CO$_2$H H··ᐟᐟ Me	Me H··ᐟᐟᐟC—OH D	CO$_2$H H··ᐟᐟᐟC—T D
$[\alpha]_D$ = +13	$[\alpha]_D$ = +13.83 (H$_2$O)	$[\alpha]_D$ = +1.6 (H$_2$O)	$[\alpha]_D$ = -81	$[\alpha]_D$ = -0.28	$[\alpha]_D$ not known!
(S)-Butan-2-ol	D-Glyceraldehyde (the simplest sugar)	L-Alanine (the simplest chiral α-amino acid)	(R)-4-Methyl-cyclohexylidene-acetic acid	(S)-(1-^{2}H)ethanol	(R)-(^{2}H,^{3}H)-acetic acid

Fig. 4.7. Molecules with a single chiral C atom (*D*, *L*, *R*, and *S* will be explained in Sections 4.6 and 4.7)

Interchange of two ligands on a single tetracoordinate chiral C atom requires bonds to be broken and reformed in all known instances. These processes are very slow at room temperature in the absence of catalysts. Alternative processes of extreme bond bending through planar or pyramidal C atoms are calculated to have even higher activation energies (probably >400 kJ mol^{-1}).

Chiral centres: other tetrahedral atoms

Molecules with tetracoordinate N^+, Si, or P as chiral centres were first studied when stereoisomerism was the only evidence for the 'shapes' of the central atoms: **4.11** and **4.12** are early examples for N and Si. **4.13**, with a chiral phosphate group, was used to study the mechanism of enzyme catalysed transfers of phosphate. The sulphone **4.14** is an example in which measurable optical activity results from oxygen isotopic substitution.

Tetracoordinate S(IV) is *not* tetrahedral because the non-bonding pair of electrons act as a *fifth 'ligand'* in a trigonal bypyramid:

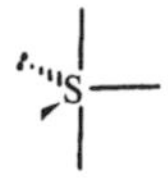

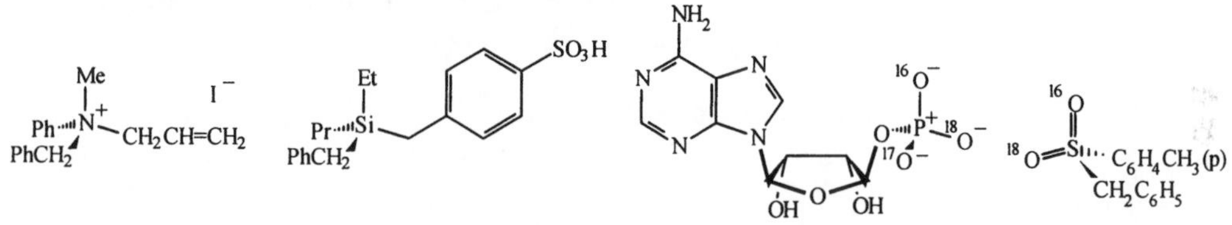

4.11 **4.12** **4.13** **4.14**: $[\alpha]_D = -0.16$

Fig. 4.8. N^+, Si, P, and S(VI) as chiral centres

Enantiomers of trialkylammonium ions $HN^+RR^1R^2$ in solution interconvert readily, except in strongly acidic solutions (Fig. 4.9).

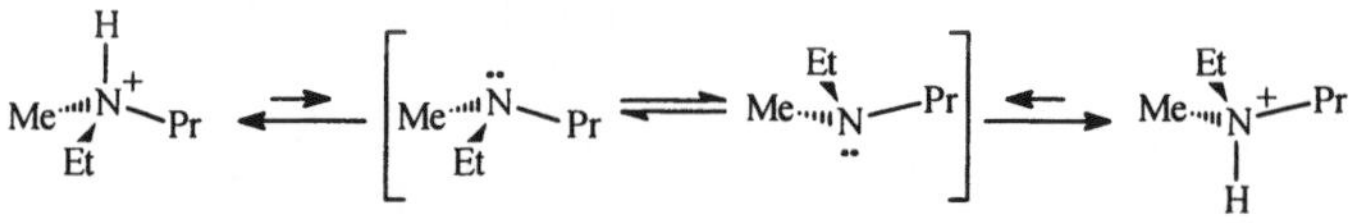

Fig. 4.9. Racemization of chiral trialkylammonium ions through the free bases

Chiral centres: tricoordinate N, P, and S

Pyramidal atoms can invert by bond bending, with a planar tricoordinate atom in the transition state. When the pyramidal atom is chiral inversion leads to racemization (in **4.19** or **4.20**) or epimerization (**4.21**). The effects of substituents on the ease of inversion are most important for nitrogen, with barriers ranging from ~0 (in amides) to >100 kJ mol^{-1} (Fig. 4.10).

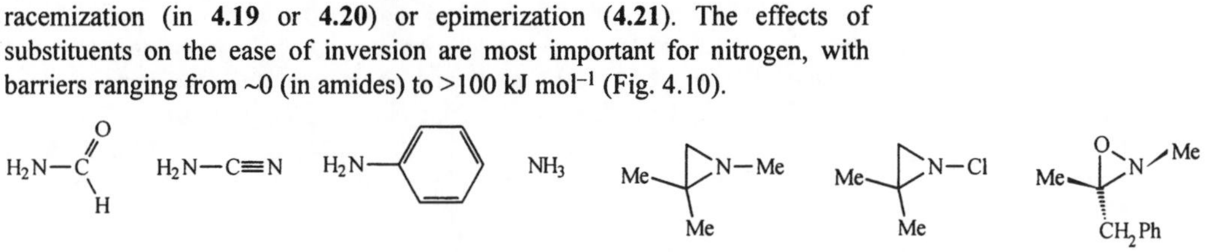

~0	6	6	22	72	112	136
4.15	**4.16**	**4.17**	**4.18**	**4.19**	**4.20**	**4.21**

Fig. 4.10. Barriers/kJ mol^{-1} to inversion at tricoordinate N atoms

The barrier to inversion is lowered by *conjugation* with π-electron withdrawing groups (in **4.15**–**4.17**), which leads to partial double bonding between C and N (see also Fig. 2.26, page 14, and Fig. 3.5, page 23). In the limit the N is *planar* as in the amide **4.15**. In contrast, ring strain and electronegative (σ-electron withdrawing) substituents *raise* the barrier to inversion. The bond angles for planar nitrogen average 120°, are much higher than the average angle, 60°, in a three membered ring (**4.19-4.21**). Electronegative substituents such as Cl (**4.20**) and O (**4.21**), and F lower the energy of the non-bonding electrons on N and make it more difficult for them to be raised to a $2p_z$ orbital in planar N.

Barriers to inversion are higher for elements in the second and later periods because the non-bonding orbitals have more *s* character: the barriers can be > 160 kJ mol⁻¹ in some phosphines. Fig. 4.11 gives early examples of homochiral compounds with pyramidal S and P.

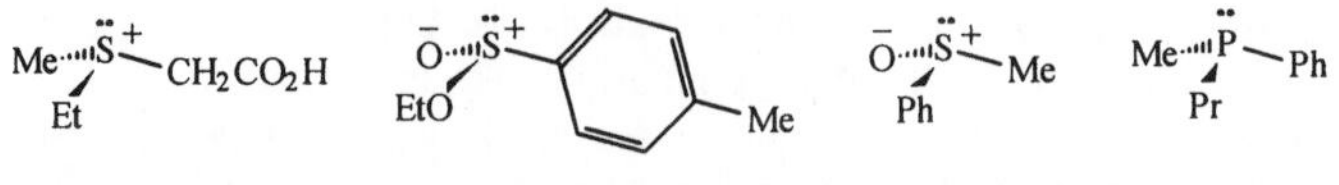

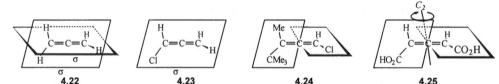

Fig. 4.11. Examples of resolved chiral compounds with tricoordinate S or P

Chiral phosphines are important as chiral ligands in catalysts used in asymmetric synthesis (see, e.g., Fig. 6.39).

Chiral axes and chiral planes

Note that molecules with chiral axes may or may not have symmetry axes. Do not confuse the two sorts of axis.

In some molecules there is a stereogenic unit that is *not* a chiral centre but a *chiral axis*. For example, in allene (propa-1,2-diene), symmetry group $\mathbf{D_{2d}}$, the >C=C=C< unit is a potential *chiral axis*. In order to eliminate the two planes of symmetry in allene (**4.22**, Fig. 4.12; the three C_2 axes are not shown) and generate a chiral molecule it is necessary that the substituents *at each end* of the C=C=C unit must be unlike as in **4.24** and **4.25**. Unlike substituents at one end, as in **4.23**, leaves one symmetry plane. The allene **4.24** has no symmetry elements but in **4.25** (glutinic acid, a natural product) one of the C_2 axes of allene (orthogonal to the 'chiral axis') is retained.

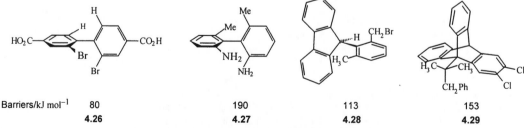

| 4.22 | 4.23 | 4.24 | 4.25 |

Fig. 4.12. Allene (propa-1,2-diene: **4.22**), an achiral monosubstituted derivative, **4.23**, and chiral derivatives **4.24** and **4.25**

In allenes the high barrier to rotation about the C-C double bonds ensures that the enantiomeric molecules do not interconvert readily, irrespective of the size of the substituents. In contrast, the *ortho* substituents in the derivatives of biphenyl **4.26** and **4.27** must prevent mutual rotation of the two benzene rings about the central bond as well as making each ring unsymmetrical. One *ortho* bromine in each ring is sufficient to allow **4.26** to be separated from its enantiomer at ~0°C. Additional *ortho* substituents as in **4.27** lead to high

Stereoisomers resulting from hindered rotation about a single bond are sometimes called *atropisomers*.

| Barriers/kJ mol⁻¹ | 80 | 190 | 113 | 153 |
| | **4.26** | **4.27** | **4.28** | **4.29** |

Fig. 4.13. Strongly hindered rotation about single bonds

barriers to rotation. High torsional barriers about sp^2-sp^3 C–C (**4.28**) and sp^3-sp^3 C–C (**4.29**) can also lead to isolable stereoisomers (see **Problems**).

Chiral planes are rather rare but one occurs in **4.6** (page 41). For purposes of nomenclature it can be reduced to a single helical unit (see page 52).

Molecules with inherently chiral skeletons

The examples of chiral molecules given so far in this chapter have involved a chiral pattern of substituents on an achiral skeleton, e.g., a tetrahedral carbon atom or an allene. Some molecules, however, have an inherently chiral skeleton and do not need substituents to make the molecule chiral. The helicenes are the clearest examples of such compounds. If [6]helicene (**4.30**) were planar the C atoms a and a' would be coincident! In the real, non-planar molecule, however, these atoms are ~ 3.03 nm apart and the strain is moderate. The very large specific rotation of [6]helicene results from the chromophore (the large aromatic system) being itself chiral.

4.30
$[\alpha]_D = -3600$ (CHCl$_3$)

4.6   Absolute and relative configurations

Dynamic symmetry in compounds: the importance of timescales

The molecules used to illustrate symmetry point groups in Section 2.10 were chosen for having a single conformer, with one exception. Such molecules become less symmetrical in many of their vibrations but the dynamic symmetry on timescales $> 10^{-10}$s is the same as that at the potential energy minimum. Most molecules are less simple and have several conformers, often with different symmetries, interconverting on fairly short timescales in most instances. Even when the symmetries are the same, as for the two chair conformers of *cis*-1,2-dimethylcyclohexane (*cis*-**3.10** in Fig. 3.13), the symmetry of the conformers, C_1, may *not* be relevant to the *compound*, which is *meso* and achiral. The symmetry of a compound, i.e., on the timescale for separations, is *never lower* than the symmetry of individual conformers. In particular the molecules of an *achiral* compound may all be chiral with a mobile equilibrium between enantiomeric molecules. Although symmetry C_1 fits the ^{13}C n.m.r. spectrum at –80°C, the spectrum at 25°C has only 4 signals of equal intensities for the 8 C atoms (see Problem 1 in Chapter 3) and has a higher symmetry. The symmetry of the latter is sometimes described as an 'averaged' symmetry but only scalars, e.g., n.m.r. chemical shifts, can be averaged. What is happening is that at 25°C pairs of nuclei are exchanging environments and *thereby averaging chemical shifts* on a timescale that is less than that of the experiment. It can be shown, however, that the permutation group for the freely ring inverting cyclohexane and the point group D_{6h} are *isomorphous*. Therefore the symmetry properties of any substituted cyclohexane that is undergoing rapid ring inversion on the experimental timescale will be reliably predicted from a *planar regular hexagonal* ring of carbon atoms for cyclohexane. The symmetry of *cis*-1,2-dimethylcyclohexane on the n.m.r. timescale at 25°C, as also on the timescales for separation, is C_s, as in formula **4.31**. This can be generalized for all cycloalkanes because the barriers to the intercoversion of their conformers are similar to that for cyclohexane.

Unfortunately the dynamic symmetry of most acyclic molecules is *not* matched by any point group and more or less arbitrary conventions have been developed for structural formulae representing *compounds*.

The idea that the 'average' of C_1 and C_1 can be, say, C_s is very odd!

4.31

The ordinary idea of a pure compound implies homogeneity to macroscopic methods of separation, which have a timescale of, say, a minute or more. Reversible first order uncatalysed 'isomerizations' (including most conformational changes), with $\Delta G^{\ddagger}/T < 0.3$ kJ mol^{-1} (i.e., $\Delta G^{\ddagger} < 90$ kJ mol^{-1} at room temperature) are too fast for separation methods to be effective.

The important quantity is $\Delta G^{\ddagger}/T$. Mixtures that are inseparable at room temperature may become separable at low temperatures, e.g., the conformers of chlorocyclohexane **4.32**, for which $\Delta G^{\ddagger} \sim 45$ kJ mol^{-1}, become separable at a very low temperature:

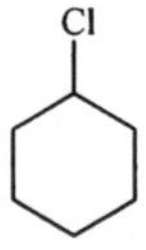

4.32

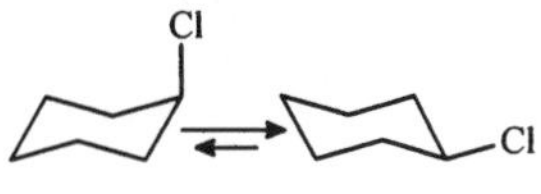

ax-**4.32**   *eq*-**4.32**

Experiment	$\Delta G^{\ddagger}/T$ /kJ mol^{-1}	Half-life
^{1}H n.m.r. at 25°C: *one* species [n.m.r., separation]	0.15	$\sim 10^{-5}$ s
^{1}H n.m.r. at –95°C: *two* species [n.m.r]	0.25	~1s
ax- and *eq*-**4.32**: separated by crystallization from CS$_2$ at –150°C	0.37	Hours

Other spectroscopic, and all diffraction, methods involve much shorter timescales.

Fischer projection formulae

Fischer devised a system for representing the configurations of sugars which is convenient for acyclic compounds with one or more chiral centres. If a molecule has a single chiral centre, e.g., *D*-glyceraldehyde (Fig. 4.14), the carbon chain is drawn vertical with the most oxidized atom at the top. By convention the vertical bonds from the *implied* central C atom go back, behind the plane of the paper, while the horizontal bonds come forward. Clearly a Fischer formula *must not* be:

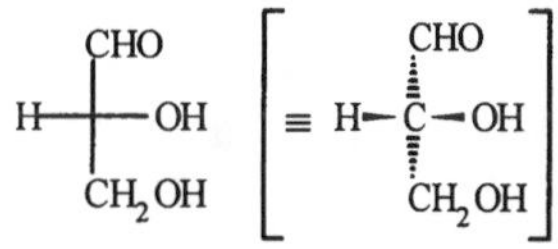

Fig. 4.14. Fischer formula for *D*-glyceraldehyde.

(a) rotated through *90° in the plane* of the paper, nor

(b) 'flipped' over, i.e., rotated through *180° out of the plane* of the paper.

A Fischer formula may be rotated through 180° *in the plane* of the paper.

Fischer formulae can be used for two or more chiral centres (Fig. 4.15) with, conventionally, *the longest carbon chain vertical*, the bonds to top and bottom atom or groups go *back* (as for a single chiral carbon atom), and all the atoms or groups to one side or the other come *forward*. If the Fischer formula has a horizontal plane of symmetry the compound is a *meso* isomer and achiral. The number of stereoisomeric compounds is obtained by enumerating all the distinguishable Fischer formulae, bearing in mind that (i) rotation through 180° in the plane of the paper and (ii) cyclic exchange of *three* ligands on a chiral C atom (at the top and bottom) are permitted.

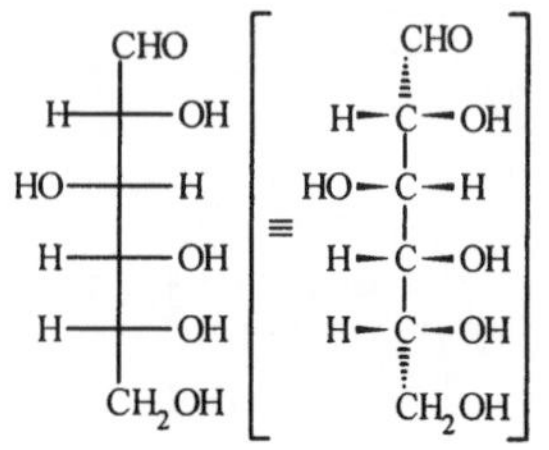

Fig. 4.15. Fischer formula for the open chain form of *D*-glucose.

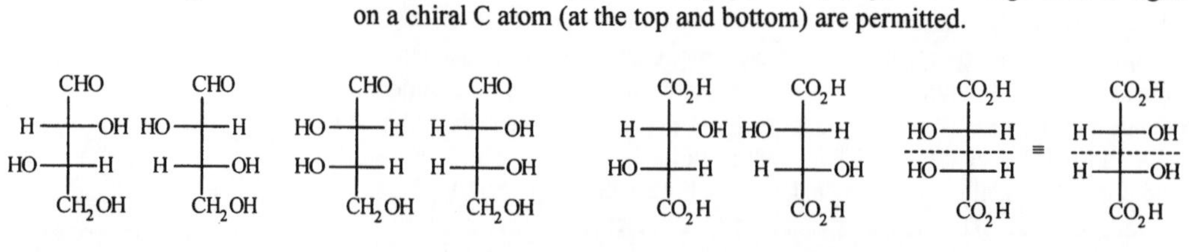

Threose **4.33**   Erythrose **4.34**   (+)- and (–)-Tartaric acid **4.35**   *meso*-Tartaric acid **4.36**

Fig. 4.16. Molecules with two chiral centres. In **4.33** and **4.34** the two chiral centres are constitutionally different and there are two pairs of enantiomers, i.e., 2×2 stereoisomers in all. **4.35** and **4.36**, however, are constitutionally symmetrical: **4.35** is a pair of enantiomers but **4.36** is a single achiral (*meso*) diastereomer (- - - - - denotes a plane of symmetry).

N.B. Fischer formulae with two or more chiral carbon atoms correspond to highly strained eclipsed conformations and will not be significantly populated but they adequately represent the dynamic symmetry of many acyclic compounds.

4.33 and **4.34** are each a pair of *enantiomers* and they cannot be inter-converted by any low energy route. In contrast **4.36** represents a *single achiral* compound in which the two halves of the molecule are related by a mirror plane *in the Fischer formula*. **4.36** was called *meso-tartaric acid* and the prefix *meso* is now used for all achiral compounds that are constitutionally symmetrical but have one or more pairs of chiral elements. Most of the conformations of **4.36** are chiral but each chiral conformation may be converted readily into its enantiomer by rotations about single bonds.

Masamune formulae

The Fischer projection was devised long before there was any information about the preferred conformations of molecules. It is now usual to follow a convention for acyclic molecules devised by Masamune. The main carbon chain is drawn as a horizontal staggered zig zag with ligands (other than H) shown with bonds coming above or going below the plane of the zig zag chain. It has the advantage of *often*, but not always, corresponding to *one* of the conformers of the molecule but this conformer may be neither the most stable nor mechanistically relevant in any given reaction. Furthermore, with branched carbon chain molecules chemists are sometimes careless in defining the 'main carbon chain', failing to follow the IUPAC rules. With constitution-ally symmetric molecules it may not be easy to spot the meso stereoisomer(s). For example, in such a molecule with two chiral centres the Masamune formula will have a centre of symmetry *i* if the chiral centres are adjacent (or separated by an even number of C atoms) but a plane of symmetry σ if they are separated by an odd number of C atoms in the main chain (Fig. 4.17).

A planar zig zag conformation for the main chain (following IUPAC nomenclature) of an acyclic molecule will not always correspond to a conformer of a molecule, e.g., when there are *syn*-1,3-substituents such as methyl or larger (see relative configurations below).

A disadvantage of the zig zag chain is that there is no correspondence between *syn* or *anti* for an acyclic molecule and *cis* or *trans* for a related cyclic compound with the same substitution pattern.

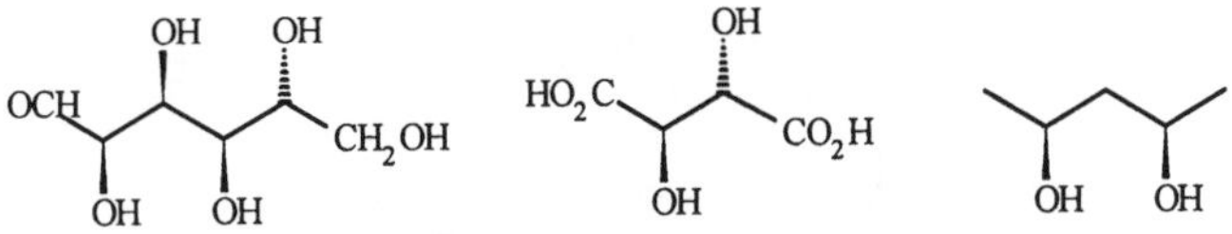

Fig. 4.17. Examples of Masamune formulae for open chain *D*-glucose, *meso*-tartaric acid, and *meso*-pentane-2,4-diol.

Monocyclic compounds

If the parent ring is drawn as a *regular planar polygon* then *cis* substituents are on the same side, *trans* substituents are on opposite sides, of the ring. Thick or wedged lines are used to indicate bonds to substituents above the plane of the ring and dotted or broken lines for bonds below the plane of the ring. This nomenclature has been adapted for use with three or more substituents. One is chosen as the reference (*r*) substituent and the others are then either *c* (*cis*) or *t* (*trans*) to it: see **4.37** for *t*-3-hydroxy-*c*-4-methyl-cyclohexane-*r*-1-carboxylic acid. The number of stereoisomeric *compounds* (distinguishable by *separation* at room temperature) with a given constitution may then be found from the number of distinguishable formulae. Individual compounds will be achiral if the formula has a plane (σ) or centre (*i*) of symmetry, otherwise they will be chiral. This is true because barriers between conformers are <90 kJ mol⁻¹ in most known compounds.

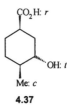

4.37

Monocyclic compounds can have structures with S_n ($n > 3$) but you are unlikely to meet examples.

The presence of a reflection symmetry element σ or *i* in such a formula does not necessarily imply that any of the *conformers* of the compound are achiral. What it does imply is that any chiral conformer can readily interconvert with its enantiomer at room temperature, as in **4.31**. Such interconversions do *not* have to go through an achiral intermediate or transition state, although many do.

Determination of absolute configurations

There is no simple method whereby the absolute configuration of chiral molecules may be derived from macroscopic properties of compounds. The first and only general method depends on 'anomalous' X-ray scattering by chiral crystals: pairs of reflections that are equally intense for an achiral crystal are unequal for a chiral crystal (this is called anomalous for historical reasons) and the signs of the inequalities can be related to the absolute configuration of the molecules. The absolute configurations of most other chiral molecules have been derived using chemical correlations. The correlations depend on reactions that either do not affect the chiral centres (or other chiral elements) or have a known stereochemical outcome. In this way *networks*, not merely chains, of relationships have been built up and the absolute configurations of almost all known chiral compounds have been determined by correlation with molecules studied by the X-ray method. Chemical reactions occasionally take an unexpected stereochemical course but errors will be detected if independent chains of evidence relating two chiral compounds are inconsistent. Examples of suitable reactions will be found in Chapter 6.

Although the quality of data from modern diffractometers is usually sufficient to determine absolute configurations of molecules of homochiral compounds, it is also vital to process the data correctly, see J.P. Glusker, M. Lewis, and M. Rossi, *Crystal structure analysis for chemists and biologists*, VCH, New York, 1994, Chapter 14.

Nomenclature for stereoisomeric molecules and compounds

Some natural chiral compounds are so important that their names are special to a single stereoisomer, e.g., cholesterol **4.39** refers to one of 256 stereoisomers with the constitution **4.38**. Other steroids are often named in a semi-systematic way as derivatives of hydrocarbons with non-systematic names and eccentric numbering, e.g., as in **4.38**, of historical origins.

4.38   **4.39**

Where a molecule has its absolute configuration implied in its name, its enantiomer may be specified by adding the prefix *ent-* to its name, as in *ent-*cholesterol.

More generally, stereochemical differences are indicated by prefixes to a chemical name that applies to all molecules with a single constitution:

a. (+) or (−), the sign of the optical rotation observed with sodium *D* light (wavelength 589 nm), is often used as a prefix to distinguish enantiomers of a compound for commercial convenience or when the absolute configuration is not known.

b. *Cis-* and *trans-* are often used for simple alkenes but *E* and *Z* are preferred: see page 52. *Cis-* and *trans-* are also used in the CIP rules and for disubstituted derivatives of monocyclic compounds (pages 28, 29, 47), and for fused rings (see Section 3.9, page 31).

The *relative* configuration of sugars is included in the name, contrary to the general rule that configuration should be specified by a prefix.

c. Some naturally occurring compounds, notably carbohydrates and amino acids, have many known stereoisomers and are so important that 'local' systems pre-dating the CIP system (below) have been retained for them.

α–Amino acids and sugars are often given the prefixes *L* and *D* (Fig. 4.18). The common α-amino acids from proteins all have the *L*-configuration. The most common sugars have the *D*-configuration, which is determined by the chiral centre *furthest* away from the aldehyde (as shown) or ketone carbonyl group.

The aminoacids and sugars have the same symbols but the two systems are quite independent.

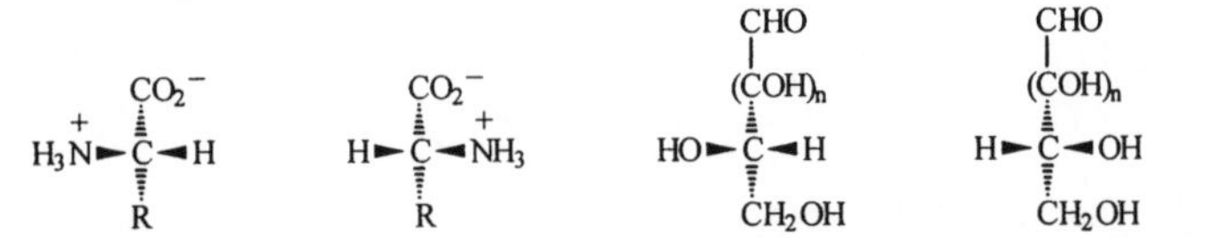

| L-α-Amino acid | D-α-Amino acid | L-sugar aldehyde | D-sugar aldehyde |

Fig. 4.18. *L* and *D* descriptors for α-amino acids and sugars

The *D* and *L* descriptors for sugars are satisfactory except for some constitutionally symmetrical derivatives.

  d. A general method for specifying the configuration of each stereogenic unit in a given molecule is provided by the Cahn-Ingold-Prelog system.

4.7 Cahn-Ingold-Prelog (CIP) system

The functions of the CIP system are (i) to assign priorities to ligands and (ii) to relate those priorities to stereochemical descriptors.

The Cahn-Ingold-Prelog system was first published in 1951 but is still being refined.

  There are two basic types of chiral stereogenic unit: a pyramidal (tripodal) unit and a non-planar chain of four atoms.

  A pyramidal molecule such as **4.40** will in practice be chiral only if X, Y, and Z are all different. The 'handedness' of **4.40** can be specified only if there are rules to assign a priority to X, Y, and Z. The most common chiral stereogenic unit in organic chemistry is a tetrahedral atom such as sp^3 carbon, as in **4.41**(W=C). The atom W is an apex common to four tripodal units with interdependent configurations. **4.41** can be reduced to a single pyramid, by ignoring the *lowest priority* ligand V, to specify the chirality of the centre W. For historical reasons we rather oddly choose to convert **4.40** into **4.41** by adding a fourth 'phantom' ligand with atomic number = 0!

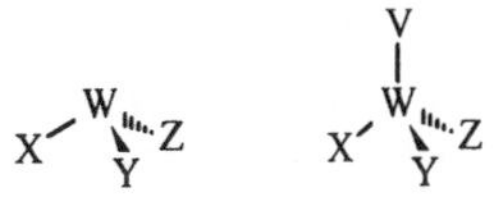

  A chain of four atoms W, X, Y, and Z joined by three bonds will be chiral if it is non-planar, irrespective of whether the atoms are alike or not. Such a chiral chain will be part of a helix that is either *P* (plus) if the torsion angle (Section 2.3) is positive or *M* (minus) if negative. Real molecules are rarely as simple as this because such a four atom chain is usually one of a bundle of chains with X—Y in common. Priority rules are needed here only to select the unique chain that defines the configuration of the stereogenic unit.

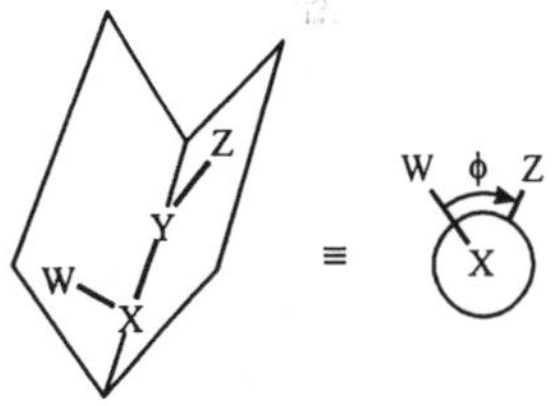

A *P* helical unit (ϕ is positive). **N.B.** X and Y may be directly bonded or may be the terminal atoms of a linear polyatomic unit, e.g., C=C=C in an allene.

Absolute and relative configurations

If one can reliably relate each chiral stereocentre in a chiral molecule to one or the other of two enantiomeric models or, equivalently, if one can assign a chirality 'descriptor' (or 'label') for each stereocentre (or other chiral unit) then the *absolute configuration* of the molecule is known. If you can only distinguish between diastereomers of a given constitution then you know the *relative configuration* of the molecule or compound.

Absolute configuration is a *topographic* property and changes on reflection. Relative configuration is a *geometric* property and is invariant to reflection.

Nomenclature for absolute configuration of a chiral centre

The absolute configuration of a 'chiral centre' (W in **4.40** and **4.41**) with three or four ligands is specified by a prefix *R* (from the Latin *rectus*) or *S (sinister)*. Denote the ligands by labels a, b, c, and, if necessary, d with the priority

$a>b>c>d$ given by the rules below. The molecule is viewed from the base a, b, c: in this way we select one tripodal unit on which to base the configuration of the tetrahedral centre. If the labels a, b, c, are in a clockwise order the configuration is R. If a, b, c are in an anticlockwise order the configuration is S:

Clockwise
order: R

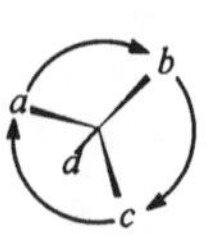

Anticlockwise
order: S

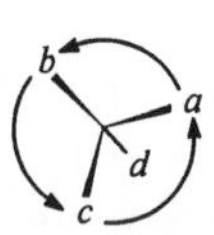

The rules apply to constitutional (rules 1. and 2.), geometric (3. and 4.), and topographic (5.) distinctions, in parallel with the distinctions between constitutional isomerism, diastereomerism, and enantiomerism made in Section 4.1.

The priorities of the groups are decided by applying the following rules in turn. The process is stopped for a ligand once it has a uniquely defined priority. Each rule is applied exhaustively before passing to the next rule. You start with the atoms directly attached to the chiral centre. If two or more of these are the same you go to the atoms two bonds from the centre and then, if necessary, to successively more remote atoms. When a ligand branches you must follow the branch of highest priority under these rules.

1. Atomic number: higher atomic number has priority over lower atomic number (irrespective of numbers of atoms): the greater number of highest priority atoms takes precedence over a lesser number. e.g., $-CHCl_2 > -CH_2Cl$. The fourth, 'empty', site on, e.g., P in a phosphine, is given an 'atomic number' of zero.

2. Isotopes: higher mass has priority over lower mass.

3. Double bond configuration: (i) Z has higher priority than E and (ii) *cis* has higher priority than *trans*, where *cis* and *trans* refer to the relationship between the substituent that includes the chiral centre and the *higher* priority group at the other end of the double bond.

Students will not meet many molecules that require rules 3 and 4. The descriptors *l*, *u*, *r*, and *s* are described later in this Section.

4. *l* (like) has a higher priority than *u* (unlike), and *r* higher than *s*.

5. Enantiomeric groups: R has higher priority than S and M is higher than P (see pages 51, 52).

The rules are easy to apply to most acyclic molecules. For example, in (+)-glyceraldehyde (see Fig. 4.19) the priority of the atoms directly attached to the chiral carbon atom is O>C>H. Therefore the OH group is a and the H is d and these are considered no further. The CHO and CH_2OH, however, can only be ordered by going to the atoms one further bond from the chiral centre, at least. The doubly bonded O in the CHO is treated as a singly bonded O atom, with a duplicate (C), and a duplicate (O) on the C, which is treated as having two O atoms attached: the duplicate atoms have no other ligands. CHO is given priority over CH_2OH, and therefore CHO is b and CH_2OH is c.

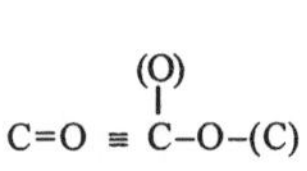

$$\underset{(O)}{C=O} \equiv C-O-(C)$$

$C\equiv N$ is treated analogously:

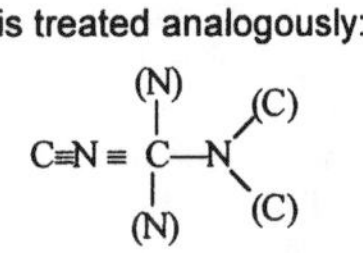

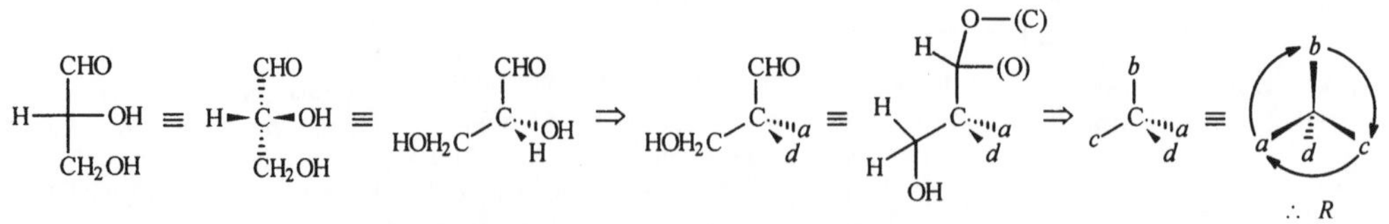

$$C\equiv N \equiv \underset{(N)}{\overset{(N)}{C}}-N\overset{(C)}{\underset{(C)}{}}$$

CHO
H—⊥—OH   ≡
CH₂OH

$$
\begin{array}{c}
CHO \\
H-\overset{\cdot}{\underset{\cdot}{C}}{\leftarrow}OH \\
CH_2OH
\end{array}
\equiv
$$

Fig. 4.19. Stages in the assignment of the absolute configuration of (+)-*D*-glyceraldehyde

Finally the chiral C atom is oriented to view a, b, c (rotation through 120° about a vertical axis), which are seen to be in a clockwise order, from the far side from d. The configuration of (+)-glyceraldehyde is therefore R.

The system is readily applied to 2 or more chiral C atoms in acyclic compounds (see problems at the end of the Chapter).

Cyclic compounds can cause difficulties. The actual structure must be converted into an acyclic 'tree' that may be different for each chiral centre. A ring is opened up by working in both directions from the initial branching point, which may be the chiral centre itself, until the branching point is reached once more and each branch is terminated with a duplicate atom (Fig. 4.20). The acid **4.42** had been regarded as an example of a molecule without a chiral centre but the acyclic tree graph shows one branch with a *cis* C=C and the other a *trans* and the third rule differentiates the two branches.

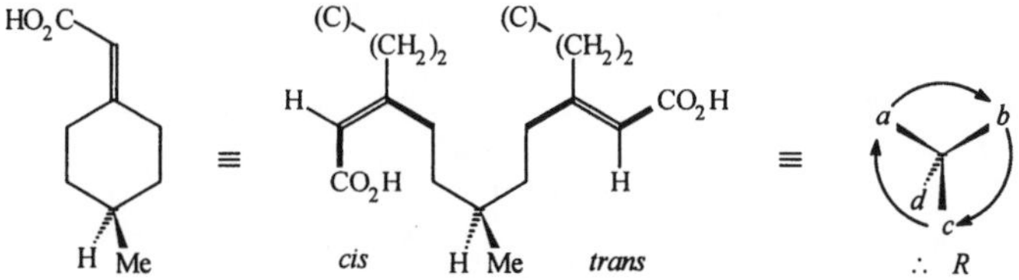

4.42

Fig. 4.20. Note the use of *cis* and *trans* to specify the relationship between the group including the chiral centre and the higher priority group ($-CO_2H$) at the other end of the C=C. *Z* and *E* cannot always be assigned within branches of a tree graph.

Groups such as phenyl, C_6H_5-, lead to complex tree graphs with multiple branching and it is easier to use tabulated priorities for many common structural units (see examples listed at the end of this Chapter).

One special case must be mentioned. In some constitutionally symmetrical molecules with an odd number of chiral carbon atoms exchange of two substituents on the central chiral carbon atom (arrowed in Fig. 4.21) may interconvert two *achiral* molecules: such a carbon atom is called *pseudochiral* (or *pseudoasymmetric*). The descriptors *r* and *s* are derived using the rules for *R* and *S*. Reflection leaves such molecules unchanged and therefore the descriptors *r* and *s* are lower case.

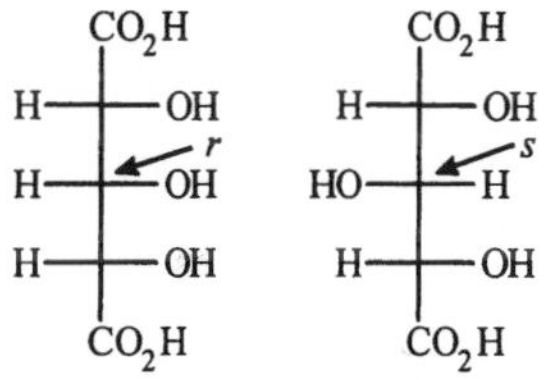

Fig. 4.21. 'Pseudochiral' centres (arrowed) in *meso* molecules

R and *S* descriptors for α-amino acids

The common L-α-amino acids are usually the (2*S*)-isomers because the COOH group usually has higher priority (*b*) than X (*c*: the second atom out from the chiral C atom is usually H, C, or a single O). In a few natural α-amino acids, however, X has the structure CH_2S-Y (e.g., Y = H in cysteine) and a single S atom has higher priority than the O atoms in COOH (Fig. 4.22).

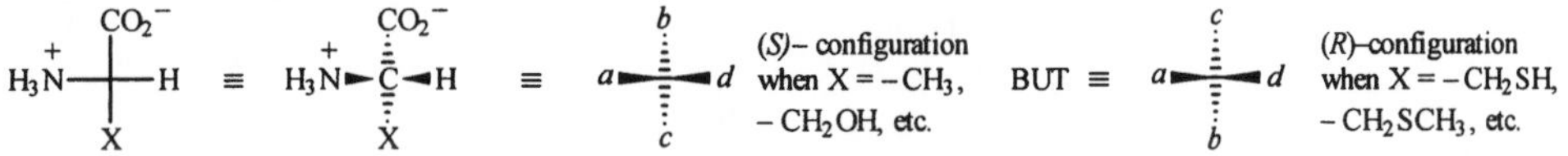

Fig. 4.22.

All the naturally occurring *L*-α-amino acids, including the sulfur containing acids such as cysteine (X = $-CH_2SH$), share common reactions with many enzymes. This simple correspondence between chemical reactions and the 'local' descriptor *L* is lost using the *R/S* system. This type of difficulty is inherent in any *general* system: there will always be some chemically significant relationships that cut across the nomenclatural differences.

Absolute configuration of chiral axes

A chiral helical unit W–X–Y–Z is commonly part of a bundle of such units with X–Y, the chiral axis, in common. For example, there are four helical units in the allene **4.24** (C=C=C is X–Y in this instance) and the highest priority unit a–X–Y–a' has the two higher priority ligands (CMe$_3$ > Me; Cl > H). If this unit has a negative torsion angle the configuration of the chiral axis is *M* (for *minus*), as in **4.43**, otherwise it is *P* (for *plus*).

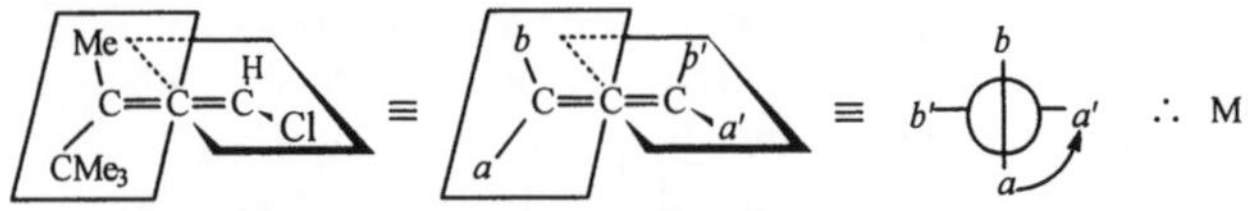

4.24   **4.43**

Fig. 4.23. Assignment of absolute configuration, *M*, to the chiral axis in **4.24**

A chiral axis with one or two *sp³* atoms requires precedence to be given to one of three ligands on each *sp³* atom. If the ligands are all different (as in **4.44**) then the priorities are determined by the CIP rules. If two are the same (as in **4.45**) then the third, unique, ligand is given the higher priority.

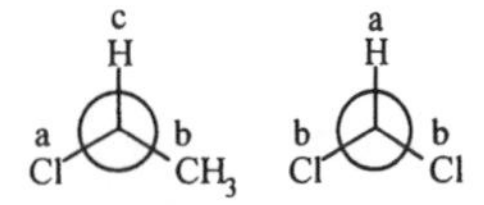

4.44   **4.45**

4.6

Fig. 4.24

This allows us to clarify the earlier names for conformations given in Fig. 3.3, page 22. The terms *synperiplanar*, etc., should refer to the relationship between the *highest* priority ligands on each of the central atoms, i.e., to the methyls in butane because the two remaining ligands on each *C*-atom are the same (Hs). If these terms are to be used for, e.g., C2–C3 in 2-methylbutane then the highest priority ligands are the methyl on C-3 and the *hydrogen* on C-2. In practice chemists usually use the terms to apply, often implicitly, to the chain of four atoms or groups that happens to be the centre of interest.

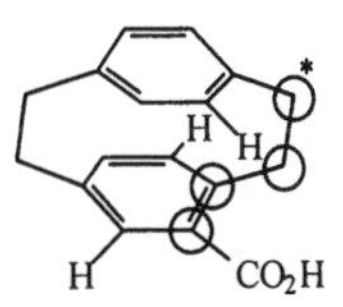

In **4.6** (Fig. 4.24), the chiral plane, the unsymmetrically substituted ring and its substituents, cannot rotate into the *potential* symmetry plane of the molecule. The chiral plane and the *next* CH$_2$ (circled and starred: Fig. 4.24) make a chiral unit that can be reduced to a single helical unit (the circled atoms constitute the highest priority helical unit), with the *M* configuration.

E and *Z* prefixes for double bond diastereomers

The prefixes *cis* and *trans* are adequate for simple double bond compounds only. Generally applicable prefixes *Z* (from the German *zusammen*, together) and *E* (*entgegung*, apart), are now in common use for double bond isomers.

The Cahn-Ingold-Prelog rules are used to assign priorities (a>b and a'>b') to the substituents at each end of the double bond. If *a* and *a'* are on the same side of the double bond then the prefix is *Z*, otherwise it is *E*. In simple alkenes such as but-2-ene *Z* corresponds with *cis* and *E* with *trans* but this is *not* general (Fig. 4.25).

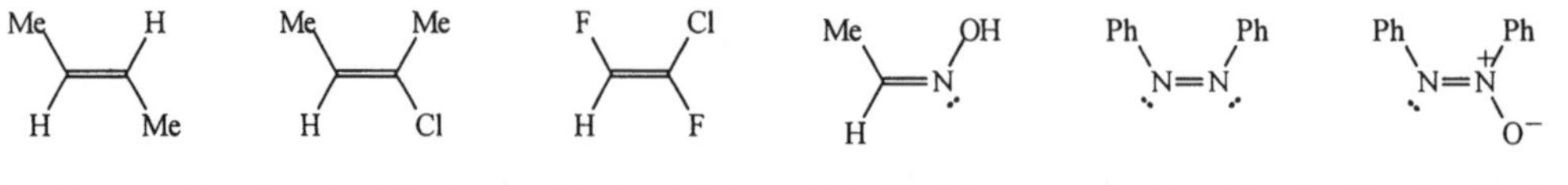

Me > H Me > H	Me > H Cl > Me	F > H Cl >F	Me > H O > :	Ph > :	Ph > :	Ph > : O > Ph
E	*E*	*Z*	*Z*	*Z*		*E*

Fig. 4.25. Examples of *Z* and *E* diastereomers

Descriptors for relative configurations

Unfortunately yet other systems for relative configurations are used by the principal abstracting services Chemical Abstracts and Beilstein.

A simple but clumsy way used in the original CIP system to express the relative configurations of chiral centres in a racemate was to give descriptors for both enantiomers, e.g., 2*R*,3*S*/2*S*,3*R* for racemic threose (Fig. 4.16). This is now shortened to 2*RS*,3*SR* but even this becomes very cumbersome when there are many chiral centres. Use of a *geometric* descriptor to indicate whether the configurations of the two chiral centres are *l* (for like, i.e., both *R* or both *S*) or *u* (for unlike) is more concise: racemic threose is *u*-2,3,4-trihydroxybutanal. When there are three or more chiral centres the centre with the lowest number (IUPAC rules) is compared with each of the others. *D*-Glucose (Fig. 4.18) has the configuration *2R,3S,4R,5R* and therefore the *l/u* descriptor for racemic glucose is *ull*.

The use of *like* and *unlike* is not limited to *R* and *S*. Stereochemical prefixes can be divided into sets of pairs (Table **4.1**) that are *like* or *unlike*.

Commas are used to separate descriptors for *different* chiral centres, e.g., *S,R* for *D*-threose. The pair of descriptors *RS* is used for a racemate with *one* chiral centre. Two pairs of descriptors imply a racemate with two like (*RS,RS*), as in erythrose, or two unlike (*RS,SR*) centres, as in threose, and so on.

Like pairs	Symbol	Unlike pairs	Symbol
RR, SS, MM, PP, RM, SP	*l*	*RS, MP, RP, SM*	*u*
ReRe, SiSi, *	*lk*	*ReSi* *	*ul*
reRe, siSi †	*Lk* †	*reSi, siRe* †	*Ul* †

Table 4.1. Like and unlike pairs of stereochemical descriptors

* See Section 4.8. †Included for completeness but not used in this book.

The descriptors *u* and *l* are concise and unambiguous but it is necessary to use the CIP rules to determine *R* and *S* descriptors for each chiral centre. Several attempts have been made to devise simpler ways to specify relative configurations for acyclic diastereomers. The prefixes *threo* and *erythro* (derived from the sugars threose and erythrose, Fig. 4.16) were first used for two chiral centres with two pairs of identical ligands (Hs and OHs in these sugars). When *the ligands* are 'similar' but not identical, however, ambiguities can, and do, arise. *Syn* and *anti* have been used for substituents on different atoms that are on the same or opposite sides of the planar zig zag main chain in Masamune formulae. If any chain atom carries two substituents then the higher priority substituent, using the CIP rules, is used to determine whether it is *syn* or *anti* to a substituent on another chain atom. This works when there is no ambiguity about assigning the main chain but in complex structures *syn* and *anti* have often been used wrongly. When *syn/anti, threo/erythro* or other 'soft' descriptors are used *the only reliable indicators of relative configurations are structural formulae showing the 3-D relationships.*

Examples of the misuse of 'soft' descriptors are given by D. Seebach et alia, *EPC Syntheses with C,C bond formation via Acetals and Enamines*, Springer, Heidelberg, 1986, pages 128-132.

Cyclic and acyclic compounds are frequently converted into one another. Unfortunately there is no regular relationship between *cis* or *trans* and *syn* or *anti* in chemically related cyclic and acyclic compounds, e.g.:

cis-3,4-dimethyl-valerolactone

anti-3,4-dimethyl-5-hydroxy-pentanoic acid

cis-2,4-dimethyl-valerolactone

syn-2,4-dimethyl-5-hydroxy-pentanoic acid

4.8 Topism

Heterotopism is the phenomenon of two or more ligands, constitutionally equivalent and therefore identical in isolation, being distinguishable when part of a molecule. For example, isotopic labelling shows that the two Hs in the

Topism is derived from the Greek τοποσ meaning a *place*.

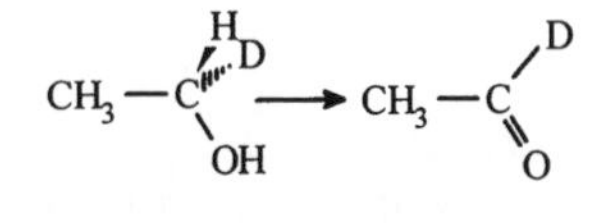

CH$_3$—C$\cdots$(H, D, OH) ⟶ CH$_3$—C(D, O)

Oxidation by nicotinamide adenine dinucleotide (NADH$^+$) catalysed by yeast alcohol dehydrogenase

CH$_2$ group of ethanol are very different in their enzyme catalysed rates of oxidation. Clearly such H atoms in ethanol are *not* identical as might be thought at first sight. A related observation is that replacement of one or the other of these H atoms by a *new* ligand (i.e., not OH or CH$_3$) leads to stereo-isomeric molecules and the C atom becomes a chiral centre. We need a suitable nomenclature for this and related phenomena.

Two (or more: see below) ligands, identical *in isolation*, may be distinguishable when part of a molecule. Two such ligands may occupy places that are *either* equivalent (*homo*topic) *or* not equivalent (*hetero*topic). The latter encompasses *constitutionally* heterotopic, such that alteration or replacement of one or the other ligand leads to constitutionally different molecules, and *stereo*heterotopic. The latter has sub-classes *enantio*topic and *diastereo*topic, such that alteration or replacement of one or other of the ligands leads to stereoisomeric molecules. Constitutional differences are so obvious that stereoheterotopic is often shortened to heterotopic.

The examples in Figs. 4.26–4.28 for simplicity all use H atoms as examples of homo- or (stereo)hetero-topic ligands but any other atoms or achiral groups could be used. *Chiral* ligands in place of the Cl atoms in, e.g., **4.46**, give more complex results that will not be considered here.

Stereoheterotopic ligands

The H atoms in each of the methylene groups in **4.46** and **4.48** (Fig. 4.26) are equivalent to one another because rotation about the C_2 symmetry axes simply interchanges the H atoms. Equivalently, replacing either H in **4.46** with a *new* ligand, i.e., not Cl, generates the same product, e.g., **4.47**. Such H atoms are *homotopic*, meaning that they occupy equivalent places in the molecule.

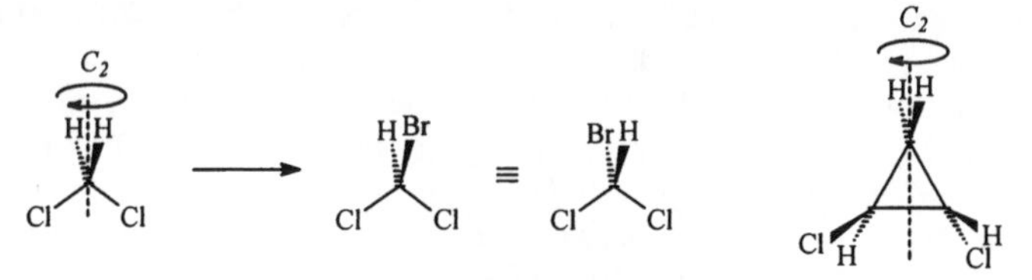

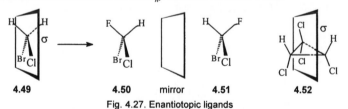

4.46   **4.47**   **4.48**

Fig. 4.26. Homotopic ligands

Purists like to use words like *enantiomeric* for *molecules* only, using *enantiomorphic*, etc., for ligands, as well as for objects in general. Most chemists use words like enantiomeric for molecules, ligands, transition states, and chemical entities generally.

The H atoms in **4.49** and **4.52** (Fig. 4.27) cannot be interchanged by a simple symmetry axis but are related by a plane of symmetry σ. The *local environments* of the two Hs are *chiral* and enantiomeric. Replacement of such H atoms by a new ligand generates enantiomeric products, e.g., **4.50** and **4.51** from **4.49**. The H atoms in **4.49** are *enantiotopic*. In Fig. 4.27 the enantiotopic H atoms are related by a σ plane but other reflection symmetry elements (the centre *i* and rotation reflection axes S_n) will have the same effect.

4.49   **4.50**   mirror   **4.51**   **4.52**

Fig. 4.27. Enantiotopic ligands

In **4.53** (Fig. 4.28) the H atoms are *constitutionally* equivalent but are in different environments that are not related by any element of symmetry. These H atoms are *diastereotopic* and replacement of one or the other by a new ligand generates diastereomeric products. There is a σ plane in **4.54** but the two H atoms are *in* the σ plane and are therefore *not* enantiotopic but diastereotopic. Replacement of the H atoms in **4.54** by a new ligand generates achiral diastereomeric (*cis* and *trans*) products.

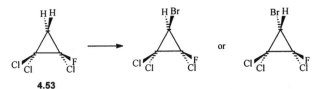

4.53

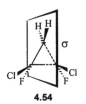

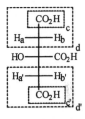

Fig. 4.28. Diastereotopic ligands

4.54

All the above examples referred to atoms as examples of homo- or heterotopic ligands but the concept and terms apply to groups of any size. Striking examples were provided by the historically important example of the enzyme controlled reactions of citric acid (**4.55**, Fig. 4.29). This Fischer formula makes it very simple to pick out the enantio- and diastereo-topic ligands. Unlike chemists and biochemists before ~1960, you should not be surprised that an enzyme catalysing the dehydration of citric acid should distinguish not only between diastereotopic H atoms, e.g., H_a and H_b but also between enantiotopic H atoms, e.g., H_a and $H_{a'}$. There is no longer a plane of symmetry relating H_a and $H_{a'}$ in the citric acid-enzyme complex and these atoms undergo reaction at *very* different rates.

4.55

Fig. 4.29. Heterotopic atoms and groups in citric acid

Stereoheterotopic faces of trigonal atoms

The idea of topism is not limited to ligands but may also be applied to *spaces* in molecules. The spaces on either *face* (see Section 2.5) of a trigonal atom, which may become occupied by an incoming atom or group, may be classified as homo-, enantio-, and diastereo-topic depending on the environment created by the ligands already attached to such an atom. In Fig. 4.30 (next page), approaches A and A' to the two faces of methanal (**4.56**) are equivalent and addition of a new ligand (i.e., neither H nor O) gives the same product **4.57** whichever face it adds to. In ethanal (**4.58**) there is a symmetry plane but *no* simple rotation axis. The environment of approach B is the mirror image of approach B' and addition of a new ligand gives enantiomeric products. Reactions at diastereotopic faces will be exemplified later (Section 6.8).

The alteration to, or replacement of, a ligand must *not* make it identical with an existing ligand at the relevant part of the molecule, throughout this discussion

There is a close parallel between classes of isomerism and of topism:

Molecules with same molecular formula			Ligands with the same structure		
Identical?	Yes →	'Homomeric' †	Related by a rotation axis?	Yes →	Homotopic
No ↓			No ↓		
Related as object and non-superposable mirror image?	Yes →	Enantiomeric	Related by a plane of symmetry?	Yes →	Enantiotopic
No ↓			No ↓		
Same constitution?	Yes →	Diastereomeric	Constitutionally equivalent site?	Yes →	Diastereotopic
No ↓			No ↓		
Constitutionally isomeric			Constitutionally heterotopic †		

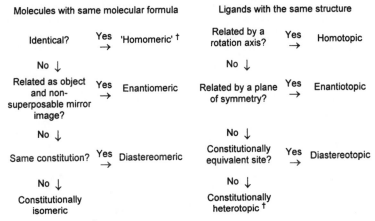

† Rarely used: too obvious!

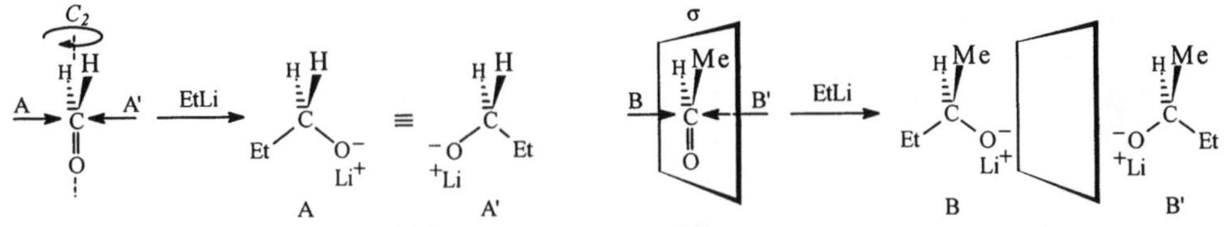

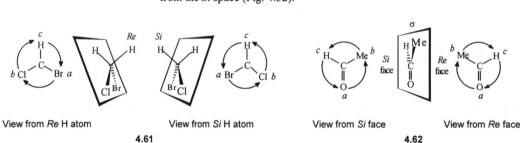

Fig. 4.30 Homotopic faces of the CO in methanal (**4.56**) and enantiotopic faces of the CO in ethanal (**4.58**)

4.9 Descriptors for heterotopic ligands and for the faces of trigonal atoms

Unfortunately there are two different systems of descriptors for heterotopic ligands in use and there is no general correspondence between them. I will describe both briefly but hope that one will soon pass into history.

Pro-R and pro-S descriptors

N.B. The *pro-R* and *pro-S* apply to the two ligands and *not* to the prochiral centre(s) to which they are attached.

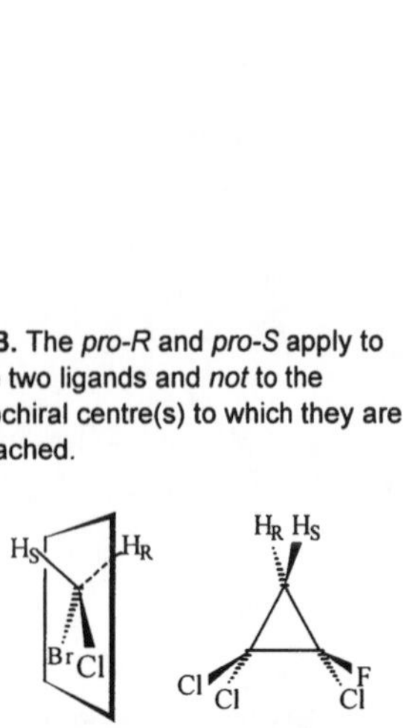

A C atom with two (stereo)heterotopic ligands (and two other different ligands) is a *prostereogenic centre*. If changing one of the ligands generates a chiral centre the C atom is *prochiral*, as in **4.59** or **4.60** (Fig. 4.31). The two similar ligands on a prochiral centre are given descriptors (labels) *pro-R* and *pro-S* (using R and S as subscripts in structural formulae where convenient, as in Fig. 4.31). A pair of heterotopic ligands have the same priority under the Cahn-Ingold-Prelog rules. Arbitrarily, let one of the heterotopic ligands (*it does not matter which*) be assigned a higher priority than the other, *without affecting the priorities of the other two ligands*. If the resulting chiral C atom is R (or S) then the ligand with raised priority is *pro-R* (or *pro-S*), otherwise it is *pro-S* (or *pro-R*) (Fig. 4.31).

Although the previous paragraph was expressed in terms of carbon centres it can be generalized to all types of stereogenic centres and axes.

Fig. 4.31. *Pro-R* and *pro-S* descriptors for enantiotopic (in **4.59**) and diastereotopic Hs (in **4.60**)

Si and *Re* nomenclature

A system for defining descriptors for prostereoisomeric centres *and* heterotopic faces of a trigonal (usually a C) atom uses the plane defined by three different ligands, with priorities $a>b>c$ determined by the Cahn-Ingold-Prelog rules, to divide space into two regions, labelled *Si* and *Re*. The order a, b, c, of the ligands is clockwise when viewed from the *Re* space, anticlockwise from the *Si* space (Fig. 4.32).

View from *Re* H atom   View from *Si* H atom   View from *Si* face   View from *Re* face

4.61   **4.62**

Fig. 4.32. *Si/Re* descriptors for H atoms in **4.61** and for the two *faces* of **4.62**

It is unfortunate that there is no general relation between *pro-S* or *pro-R* and

Si or *Re* for heterotopic atoms or groups. If the heterotopic atoms or groups are the lowest priority (as in **4.61**) or highest priority then *pro-S* (*pro-R*) matches *Si* (*Re*) but this is not true when such groups are of intermediate priority, e.g., the enantiotopic Me groups in propan-2-ol (**4.63**).

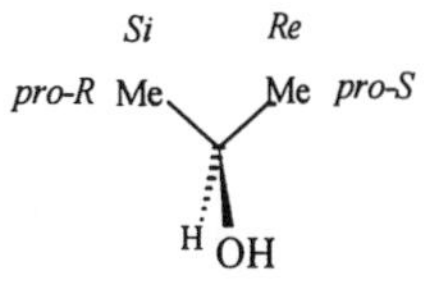

Si   *Re*

pro-R Me   Me *pro-S*

H OH

4.63

4.10 Topism and timescales

The examples used in Figs. 4.26–4.32 were chosen for simplicity in the sense that the relationships were independent of the experimental method used to study them. In the majority of molecules, however, the observed relationships depend on the *timescale* of the experiment, as in stereoisomerism (Section 4.6), and it is now necessary to look more closely at homo- and hetero-topic atoms and groups.

In the favoured *staggered* conformation of methylamine (**4.64**; symmetry C_s) it is clear that the three methyl H atoms are not in identical environments at any moment. The difference between H_t and H_{g_+} or H_{g_-} shows up in, e.g., the stretching frequencies (infrared spectra) and bond lengths (from *ab initio* MO calculations: the difference agrees with experimental correlation of bond lengths and vibration frequencies) of the C–H bonds. The absorption of electromagnetic radiation is fast for both rotational and vibrational spectroscopy relative to conformational changes and diastereotopic H atoms may be distinguished in many instances.

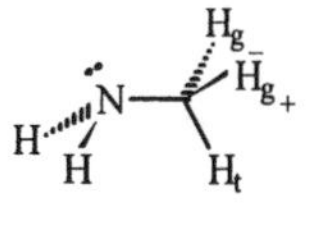

Although chemical shifts and coupling constants are sensitive to environment, the three C-H protons in methylamine exhibit *identical* chemical shifts and coupling constants (with, e.g., ^{13}C) in the ^1H n.m.r. spectrum. This is because the ^1H resonance frequencies would differ by a few hundred Hz at most and would require a correspondingly slow rotation of the methyl group. Thus the *average* environment of the three protons is the same so far as n.m.r. is concerned.

Both internal rotation about the single bond and N inversion are very fast in MeNH$_2$ but barriers to both processes can be much higher in more complex molecules (see, e.g., Fig. 3.5, page 23; Fig. 4.10, page 43; and Fig. 4.13, page 44).

Further examples of topism are given in the problems.

H N H_{g_-} H_{g_+} H H_t

4.64

Bond lengths (in pm):
$r(C-H_g) = 108.9$; $r(C-H_t) = 108.0$
Stretching 'frequencies' (in cm^{-1}):
$v(C-H_g) = 2880$; $v(C-H_t) = 2955$

Fig. 4.33. Stereoheterotopic C–H bonds in methylamine

Problems

1. Assign *R*, *S* descriptors to the chiral centres depicted in **4.11**, **4.12**, **4.14**, **4.21**, **4.37**, and **4.39**, and to the four tricoordinate atoms in Fig. 4.11.

2. Assign *P*, *M* descriptors to the chiral axes in **4.25–4.27**.

3. Assign descriptors to all the heterotopic atoms and groups in citric acid, using the Fischer formula from Fig. 4.29.

4. The acid **4.65** reacts with *racemic* PhCHMeNH$_2$ to form four diastereomeric amides. What are the *l*/*u* descriptors and symmetries of these diastereomers?

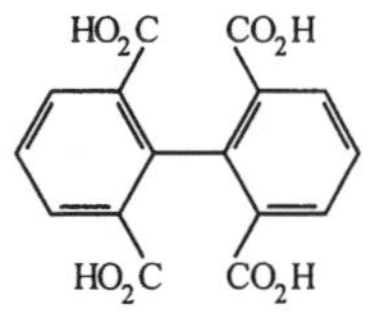

5. Enumerate the chiral and achiral diastereomers of **4.66** ('truxillic acid'), and **4.67** ('truxinic acid')

6. Identify the diastereomer of cyclopentane-1,2,3,4-tetracarboxylic acid that is chiral, has five resonances in its ^{13}C n.m.r. spectrum, and forms a bis-anhydride.

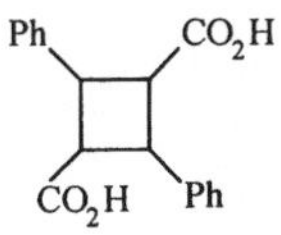

HO$_2$C   CO$_2$H

HO$_2$C   CO$_2$H

4.65

Ph   CO$_2$H   CO$_2$H   Ph

CO$_2$H   Ph   CO$_2$H   Ph

4.66 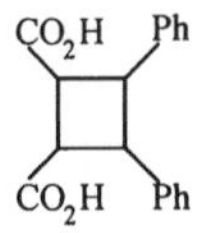   **4.67**

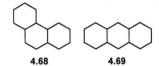

4.68   **4.69**

7. (a) How many diastereomers are there of perhydrophenanthrene (**4.68**) and perhydroanthracene (**4.69**)? (b) Which, if any, are *meso*? (c) Which, if any, cannot exist in chair-chair-chair conformers?

8. Explain the following:

(a) The Hs in the CH_2 groups in $(PhCH_2)_2NMe$ have the same chemical shift at 185K but different chemical shifts at 122K. Similar changes take place in its *salts* in H_2O between pH 2.7 and pH ~0.

(b) 1,2,3,4-Tetramethylcycloocta-1,3,5,7-tetraene is a chiral compound ($\Delta G^\ddagger$ = 133 kJ mol^{-1} for racemization).

Further reading

The CIP system: (a) R.S. Cahn, C. Ingold, and V. Prelog, *Angew. Chem. Int. Ed. Engl.*, 1966, **5**, 385; (b) V. Prelog and G. Helmchen, *Angew. Chem. Int. Ed. Engl.*, 1982, **13**, 263.

Topism nomenclature: K. Mislow and M. Raban, *Top. Stereochem.*, 1967, **1**, 1; H. Hirschmann and K.R. Hanson, *J. Org. Chem.*, 1971, **36**, 3293.

Chiroptical properties: G. Snatzke, in *Methodicum Chemicum*, ed. F. Korte, Academic, New York, 1974, Chapter 5.7; P. Crabbé, *ORD and CD in Chemistry and Biochemistry: An Introduction*, Academic, New York, 1972.

CIP priority of a selection of ligands (in ascending order from n =1 to n = 42)

n		n		n	
1	H	15	s-Bu	29	$-CPh_3$
2	Me	16	1-propenyl	30	$-C_6H_4NO_2$-o
3	Et	17	t-Bu	31	$-NH_2$
4	Pr	18	2-propenyl	32	$-NH_3^+$
5	Bu	19	$-CCH$	33	$-NHCOOCH_2Ph$
6	pentyl	20	Ph	34	$-NMe_3^+$
7	hexyl	21	$-C_6H_4Me$-p	35	$-N=O$
8	3-Me-Bu	22	$-C_6H_4NO_2$-p	36	$-NO_2$
9	2-Me-Pr	23	$-C_6H_4Me$-m	37	$-OSOPh$
10	allyl	24	$-C_6H_3Me_2$-3,5	38	F
11	$-CH_2t$-Bu	25	$-C_6H_4NO_2$-m	39	$-SH$
12	$-CH_2C\equiv CH$	26	$-CCMe$	40	Cl
13	$-CH_2Ph$	27	$-C_6H_4Me$-o	41	Br
14	i-Pr	28	$-C_6H_3Me_2$-2,6	42	I

5 Racemates and resolution

In this chapter I enlarge the concept of stereoisomerism to groups of molecules, e.g., in crystals, or a molecule surrounded by a solvent or adsorbed on a surface. These are essential in some methods for *resolving* racemic compounds, i.e., separating mixtures of enantiomeric compounds.

Scalar properties such as interatomic distances and energies are independent of the absolute configuration of a chiral molecule that is either isolated or in an achiral environment. This is no longer true if the chiral molecule is put into a *chiral* environment. Enantiomeric chiral molecules interact differently with other chiral molecules because such interactions are *diastereomeric*. Examples of chiral environments are chiral solvents or chiral stationary phases in chromatography (Section 5.5, pages 64, 65), enzymes (Section 6.8) or, simply, homomers (see Fig. 4.3, page 37) and enantiomers of the chiral molecule itself. This last possibility appears in the difference in properties of crystals of homochiral and racemic chiral compounds.

5.1   Properties of racemates

If a chiral compound is synthesized from achiral or racemic reactants (and catalysts, etc.) then it will be formed as a *racemate*, a (statistically) 1:1 mixture of enantiomeric compounds (page 78). Racemates may also result in reactions that (i) allow enantiomers to interconvert (*racemization*) or (ii) go through achiral transition states or intermediates. Enantiomeric compounds have identical vapour pressures, solubilities, etc. and therefore racemates, 1:1 mixtures, cannot be separated by the methods used for ordinary mixtures. *All* methods for separating racemates depend on *diastereomeric* interactions. There are few generalisations to guide the choice of method used. If a single enantiomer is required resolution will usually waste at least half of a racemic mixture. Resolution is therefore often wasteful as well as uncertain and it is usually better to generate pure enantiomers directly through enantioselective synthesis if this is possible (see Section 6.8, pages 78–83).

Not only is it difficult to separate racemates but it can be difficult to determine how well a separation has succeeded (see Section 5.5, pages 64, 65).

There is a small difference in the interactions between two like chiral molecules, i.e., R,R and S,S pairs, and between enantiomeric molecules, i.e., R,S pairs. When *liquid* enantiomeric compounds are mixed there are small changes, typically ~200 J mol^{-1} or less, in heat content that will cause, at most, very small changes in vapour pressure. In crystals, where the lattice controls the relative positions of neighbouring molecules, these differences in heat content and related properties can be much larger and may lead to useful differences. In practice most racemates form *achiral* crystals with a 1:1 ratio of enantiomers (*racemic compounds*), ~10% give a mixture of chiral crystals, each containing a single enantiomer (*racemic conglomerates*), and a small number form crystals containing both enantiomers distributed more or less randomly (*racemic solid solutions*).

If a chiral compound has two or more *chiral* centres (or other stereogenic elements) *all* of them must change in order to get to the enantiomer. This is often overlooked by students. *Epimerization* is a change at only one of two or more chiral centres, generating a *diastereomer* of the reactant.

Waste of the unwanted enantiomer may be avoided if it is possible to racemize it, particularly if this can be done under the conditions used in a resolution (see next page).

Mixing enantiomers always leads to an increase in *entropy*. Racemization in dilute solutions is therefore always thermodynamically favourable.

Crystal structures of racemates

When a mixture of enantiomeric molecules crystallizes there are three limiting types of behaviour. If enantiomeric molecules prefer to crystallize together in an ordered achiral crystal (a *racemic compound*) the latter may be higher melting (as in tartaric acid) or lower melting (as in Fig. 5.1(a)) than either enantiomer. If enantiomeric molecules co-crystallize with little or no preference as to which is next to which a *racemic solid solution* is formed: this is rare. Racemic solid solutions may be lower melting (very rare) or higher melting (as in Fig. 5.1(b)) than the enantiomers, or the melting point may be (almost) the same for all compositions. Racemic compounds and racemic solid solutions cannot be separated by crystallisation.

> Some racemates show more complex behaviour, e.g., (i) temperature induced changes between conglomerates and racemic compounds and (ii) solid solutions for a *limited* range of compositions.

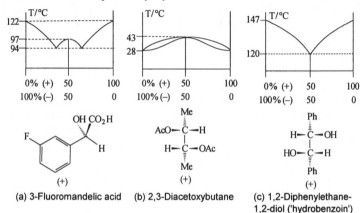

(a) 3-Fluoromandelic acid   (b) 2,3-Diacetoxybutane   (c) 1,2-Diphenylethane-1,2-diol ('hydrobenzoin')

Fig. 5.1. Melting points of mixtures of enantiomers. (a) a racemic compound, (b) a racemic solid solution, (c) a racemic conglomerate (simple eutectic system).

> A single crystal weighing ~10^{-5}g of the heterocyclic helicene **5.1** (an example of Fig. 5.1(c)) was used to determine the absolute configuration by X-ray diffraction and then dissolved in order to measure the optical rotation.

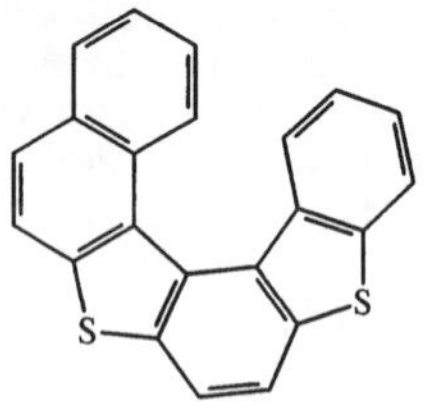

5.1

In the third possibility (Fig. 5.1(c)) identical molecules interact better with one another than with enantiomeric molecules. In this case enantiomeric compounds in a racemate crystallize separately in enantiomeric chiral crystals, giving a simple eutectic mixture, a *racemic conglomerate* (preferred term: sometimes *racemic mixture* is used). If the enantiomeric crystals are visibly different then they can be separated by hand sorting (see Section 1.1 for the first example); if not, a single crystal may be enough for a number of experiments, e.g., the helicene **5.1**! An effective variant, that has been used on a commercial scale, is to seed a supersaturated solution of a racemic conglomerate alternately with crystals of one or the other enantiomeric form. Unfortunately the formation of a conglomerate, as in (c), has been observed for fewer than 300 pairs of enantiomers.

A special case arises when enantiomers interconvert at an appreciable rate in solution or melt, spontaneously or with catalysis. It is then possible to seed with crystals of one enantiomer and slow crystallization will then allow the other enantiomer to racemize and maintain (approximate) equality of

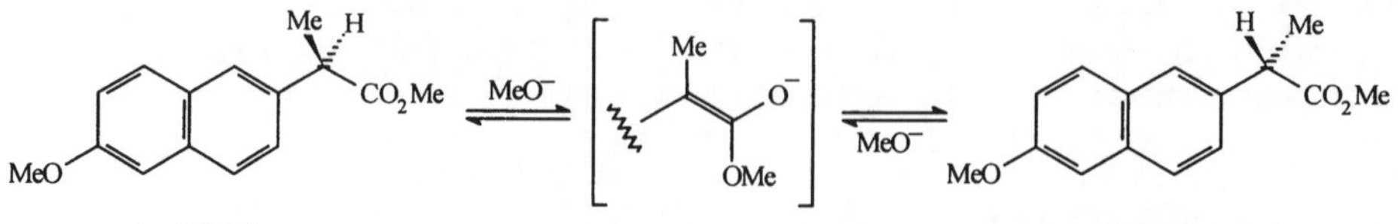

(*R*)-**5.2**   (*S*)-**5.2**

Fig. 5.2. The resolution of of the methyl ester **5.2** of naproxen, an analgesic, through NaOMe catalysed interconversion of enantiomers in a supercooled melt and selective seeding to induce crystallization of *S*-**5.2** (87% yield).

concentrations of the enantiomers in the melt (Fig. 5.2). In other readily racemized systems it is possible for one enantiomer to crystallizes first and the whole sample may then form crystals of one handedness spontaneously!

5.2   Resolution through diastereomeric derivatives

If the enantiomeric components of a racemate can be converted into a mixture of diastereomeric derivatives, from which the enantiomers of the original compound can be recovered eventually, using an enantiomerically pure chiral reagent, then the problem of separating a racemate is changed to separating *diastereomeric* compounds, usually by crystallization or chromatographically.

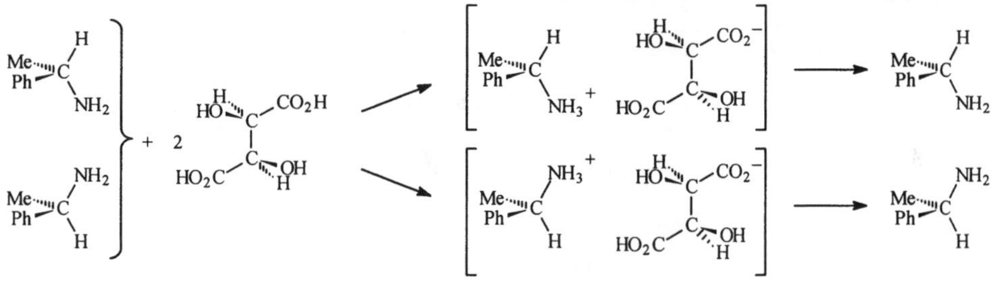

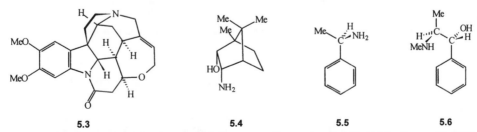

Fig. 5.3. A typical resolution through diastereomeric derivatives: resolution of 1-phenylethylamine with (2R,3R)-(+)-tartaric acid

Racemates that are acids or bases are commonly resolved by formation of diastereomeric salts with a chiral base or acid respectively (Fig. 5.3). It is easy to make salts and easy to recover acids and bases from them. A small difference in the packing of ions can have a large differential effect on the lattice energies and thence on the solubilities of such salts. It is the less soluble of the diastereomeric salts that is the easier to purify and this makes one enantiomer much easier to isolate pure. If both enantiomers of the resolving agent are available, however, alternation in their use can lead to resolutions giving both enantiomeric compounds in good yield and high e.e. (Marckwald's principle). Natural and semi-synthetic chiral acids and bases are usually available in one enantiomeric form only but an increasing number of synthetic agents are now made commercially in both enantiomeric forms (Fig. 5.4).

If you want to resolve a *neutral* racemate one tactic is to convert it into an *acidic* derivative (*basic* derivatives are uncommon) and resolve the latter with

There are few guidelines in the art of resolution and this uncertainty in the outcome is one good reason for avoiding resolution if possible.

e.e. = enantiomeric excess: see Section 4.3.

'Semi-synthetic' refers to reagents prepared from naturally occurring compounds, e.g., **5.8** from (+)-tartaric acid, **5.7**.

5.3	**5.4**	**5.5**	**5.6**

Fig. 5.4. Chiral bases used for resolving acids. Brucine (**5.3**) is a naturally occurring alkaloid which is often used to make salts of acids but which also forms *inclusion* compounds (see pages 63, 64) with many neutral compounds. The aminoborneol **5.4** (from camphor) is a 'semi-synthetic' compound available in one enantiomeric form. The amines **5.5** and **5.6** ((1R,2S)-ephedrine) are available as shown and in the enantiomeric forms.

a chiral base. For example, a racemic alcohol is treated with a cyclic acid anhydride (most often phthalic anhydride) to give an acidic *mono-ester* of a dibasic acid. This acid ester can be resolved with a chiral base such as brucine

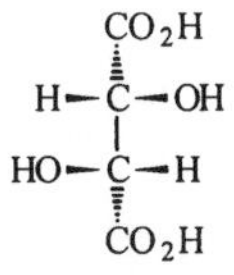

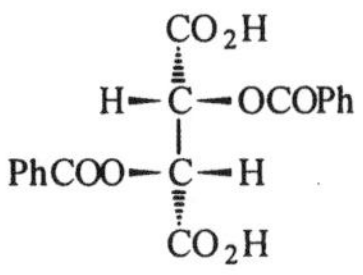

Fig. 5.5. Acids used to resolve bases. The naturally occurring (+)-tartaric acid **5.7** is much more readily available than its enantiomer and is the starting material for a number of useful resolving agents such as **5.8** (for resolving bases) and **5.15** (see Fig. 5.6) used to resolve ketones and aldehydes. The acids **5.9–5.11** are available in both enantiomeric forms. **5.11** and **5.12** are stronger acids than carboxylic acids and this can be helpful in preparing salts of weak bases.

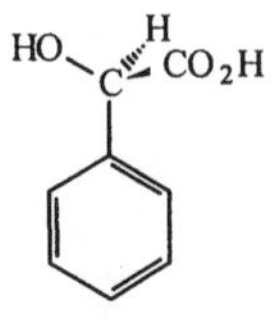

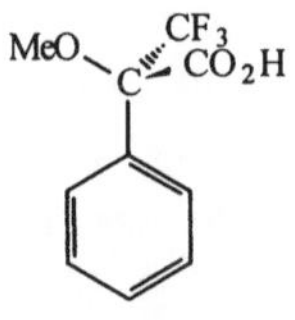

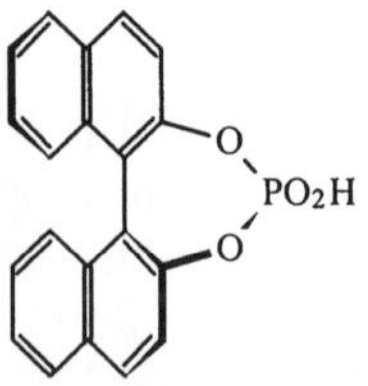

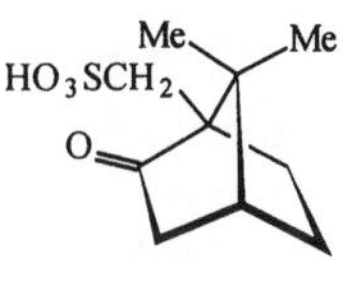

5.2, and the separated enantiomeric acid-esters can be hydrolysed or reduced to get the enantiomeric alcohols. At one time this was very widely used but the efficiency of chromatographic separations has made it more common to use covalent derivatives, e.g., esters of alcohols, which can be separated by chromatography or crystallization or a combination of these methods.

Figure 5.6 gives a selection of reagents for the resolution of racemic alcohols and aldehydes and ketones.

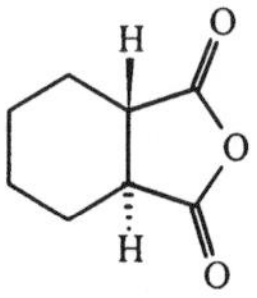

Fig. 5.6. Reagents used in resolution of racemic alcohols and aldehydes and ketones. **5.13** forms acid esters while **5.14** gives urethanes (RNHCOOR') with alcohols. Racemic aldehydes and ketones react with **5.15** to give diastereomeric acetals and ketals, while **5.16** is a chiral analogue of semicarbazide ($H_2NCONHNH_2$) for forming derivatives of aldehydes and ketones.

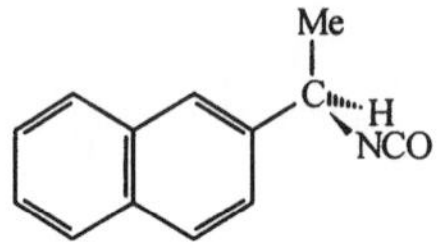

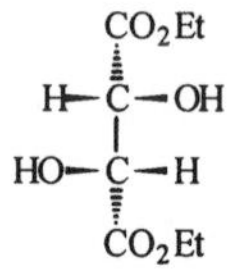

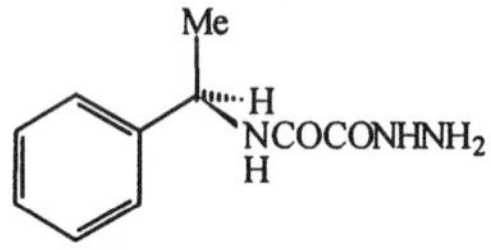

The resolution of racemic *trans*-cyclooctene (Fig. 5.7) is an example of the use of a metal centred complex to form diastereomeric derivatives with a racemic alkene. The twisted C=C in *trans*-cyclooctene coordinates more strongly than ethene with platinum in the square planar complex but after the separation of the diastereomeric complexes it is displaced in turn by CN^-.

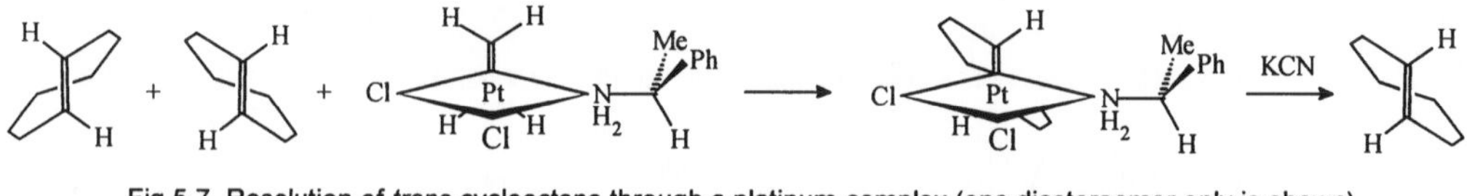

Fig 5.7. Resolution of *trans*-cyclooctene through a platinum complex (one diastereomer only is shown)

5.3 Resolution using chiral chromatographic stationary phases

The interactions of enantiomeric molecules with a chiral solvent, another chiral solute, or with the surface of a chiral adsorbent are *diastereomeric*. Differences in such interactions for a pair of enantiomers are generally small and may be effective only if applied repeatedly, as in chromatography. Useful solute-surface and solute-solvent interactions for chromatographic separations were first observed for pairs of compounds forming at least one strong hydrogen bond. For example, 'Troeger's base' (**5.17**: Fig. 5.8), which is readily decomposed by acids, was first resolved by chromatography on powdered lactose (**5.18**). The most widely used solid chiral adsorbents appear to be derivatives of natural polymers, e.g., **5.19** but fully synthetic chiral polymers such as **5.20** are very effective.

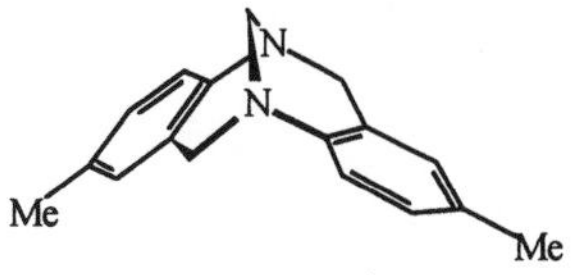

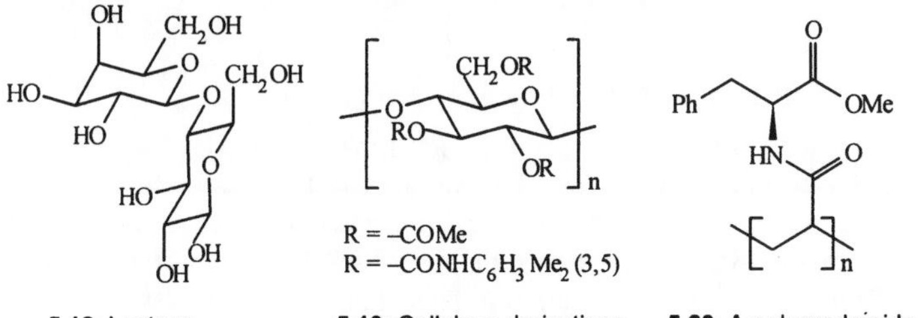

5.17: Troeger's base

5.18: Lactose   **5.19**: Cellulose derivatives   **5.20**: A polyacrylamide

$R = -COMe$
$R = -CONHC_6H_3Me_2 (3,5)$

Fig. 5.8. Chiral stationary phases for preparative chromatographic resolutions

5.4   Resolutions using complexes and inclusion compounds

It is difficult to convert aromatic hydrocarbons into diastereomeric derivatives for resolution but many form quite strong complexes with polynitroaromatic compounds, e.g., **5.21**. In most instances resolution results because one

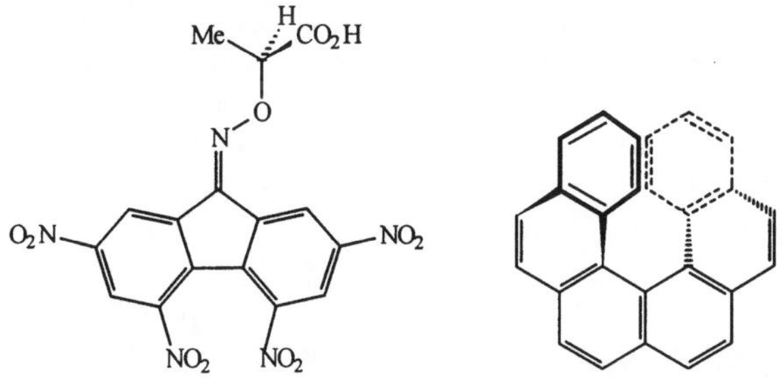

(*S*)-**5.21**   (*M*)-**5.22**: (*M*)-Hexahelicene

Fig. 5.9. Unusually, (*S*)-**5.21** forms a relatively stable *soluble* complex with (*P*)-**5.22**, so that (*M*)-**5.22** crystallizes when (*S*)-**5.21** is used to resolve racemic **5.22**

enantiomer of the hydrocarbon forms a sparingly soluble complex but Fig. 5.9 gives an instance in which the complex of one enantiomer stays in solution while the other enantiomer crystallizes.

A method of resolution applicable to many classes of compound but of particular value for inert or otherwise difficult compounds makes use of *inclusion* compounds. Some molecules cannot pack efficiently into crystals without other molecules to fill up holes or channels. Some of these inclusion compounds have strong intermolecular hydrogen bonds but appropriate *shapes* are often the most important factor. Brucine (**5.3**, Fig. 5.4), which forms crystal compounds with a wide variety of small molecules, may have been the first to be used in this way but the phenomenon was first recognised in tri-*o*-thymotide (**5.23**). Examples of compounds that have been (partly) resolved in this way are given in Fig. 5.10.

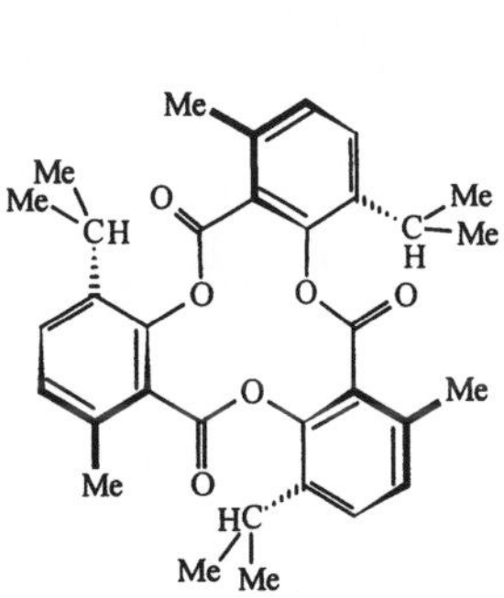

5.23

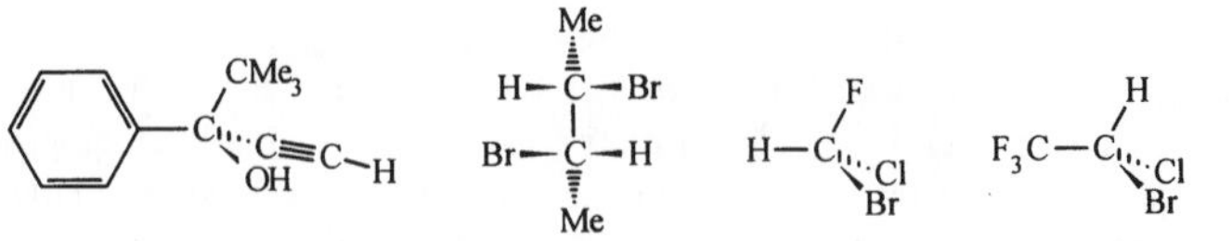

Fig. 5.10. Compounds partly resolved through forming inclusion compounds with brucine (**5.3**). **5.24** is the anaesthetic halothane.

5.5   Determination of enantiomeric excess

Measurements of optical rotations [α] are sensitive to achiral as well as chiral impurities and errors can easily reach 2%.

In Section 4.3 it was shown that if the specific rotation [α] is known for a pure enantiomer previously prepared then the enantiomeric excess of another sample may be estimated from its specific rotation $|[\alpha]| \leq |[\alpha]|_{(100\%)}$. But how do we know that the earlier sample *was* pure? When repeated purification, preferably by more than one method, fails to increase [α] it is usually assumed that the compound is enantiomerically pure. Unfortunately this is laborious, not completely reliable, and not sufficiently sensitive if we need to determine the amount of the minor enantiomer when the e.e is near 100% (see Fig. 5.12).

The chiral reagent must react quantitatively with the two enantiomers, otherwise there will be selectivity in the formation of the diastereomeric derivatives (see Fig. 6.50, page 81, and related discussion).

Methods that depend on separate measurement of each enantiomer, or a derivative, in a mixture are better in principle. This is possible if enantiomers can be either (i) converted quantitatively by a homochiral reagent into diastereomeric derivatives for conventional analysis (spectroscopy or chromatography) or (ii) differentiated in a chiral environment.

An example of the first type of method is provided by Mosher's reagent, **5.25** (the acid chloride of the resolving agent **5.10**, Fig. 5.5), which forms derivatives with many amines and alcohols. The diastereomeric derivatives can be separated by chromatography or detected and estimated by ^{19}F n.m.r. The greatest problem is that the chiral reagent must have a high enantiomeric excess (Fig. 5.11).

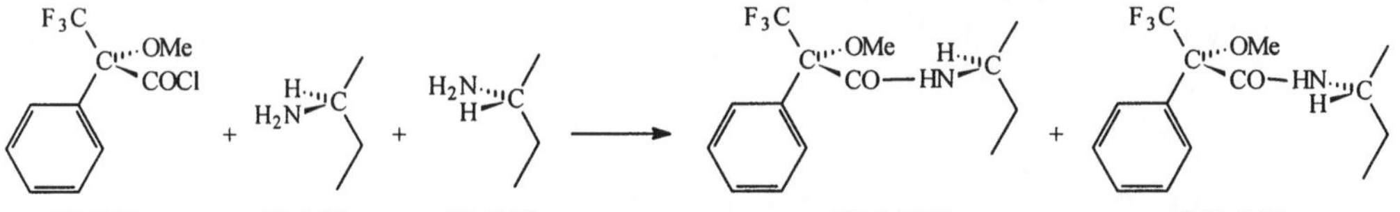

(R)-**5.25**   (S)-**5.26**   (R)-**5.26**   (S,S)-**5.27**   (S,R)-**5.27**

[(S)-**5.25** will give rise to: (R,R)-**5.27**   (R,S)-**5.27**]

Fig. 5.11. Determination of enantiomeric excess using a chiral reagent (R)-**5.25** to convert amines (S)- and (R)-**5.26** into diastereomeric derivatives (S,S)- and (S,R)-**5.27**. If the reagent is not enantiomerically pure then the enantiomer (S)-**5.25** will generate (R,R)- and (R,S)-**5.27** which will be analytically indistinguishable from (S,S)- and (S,R)-**5.27** and the *apparent* enantiomeric purity of **5.26** will be less than the true value.

In the second method the chiral environment is usually either a chiral chromatographic stationary phase used in very efficient analytical columns, or a chiral solvent or complexing agent in n.m.r. Gas chromatography using a chiral stationary phase can achieve a sensitivity of 0.01% or better in detecting a minor enantiomeric component. This sensitivity is required in some enzyme catalysed reactions, e.g., the conversion of fumarate **5.28** to *L*-aspartate **5.29** (Fig. 5.12). Aspartate, an anion, must be converted into a volatile covalent derivative before analysis. There is no likelihood of selectivity in making derivatives with achiral reagents (see Fig. 6.44, page 80), unlike diastereomeric derivatives prepared with *chiral* reagents (see Fig 6.50, page 81).

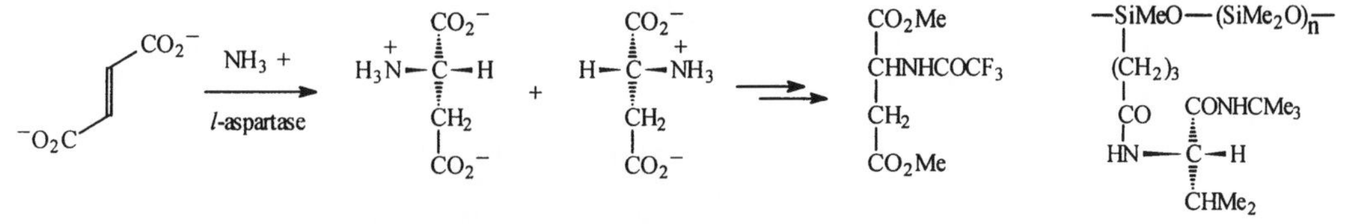

5.28 **5.29**: ~99.99% ~0.01%

Enzymatic reaction in H_2O Volatile derivative Chirasil-Val

Fig. 5.12. The amination of fumarate to (*L*)-aspartate is extremely enantioselective. The high sensitivity needed to measure the *D*-isomer is achieved by conversion to a volatile covalent derivative for analysis by gas chromatography using the chiral stationary phase Chirasil-Val (derived from valine, an α-aminoacid) and a mass spectrometer as a selective detector.

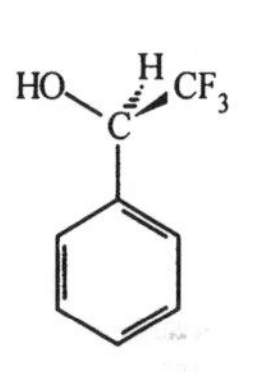

(*R*)-5.30

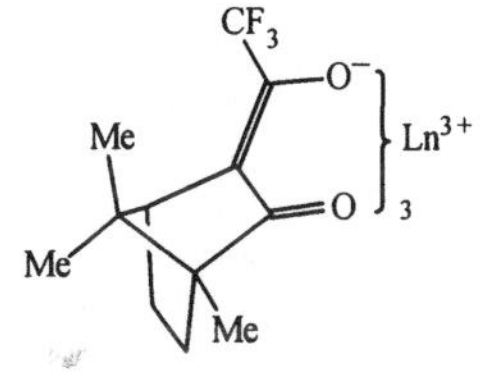

5.31

Much simpler, but less general, is to use *either* a chiral *solvent or* a chiral *shift reagent* in n.m.r. 1-Phenyl-2,2,2-trifluoroethanol **5.30** is a typical chiral solvent for this purpose. The –OH group allows it to hydrogen bond to suitable solutes, notably alcohols and amines. The effect of the hydrogen bonding is to limit the relative orientations of solvent-solute pairs and thereby minimise the extent to which perturbations of the chemical shifts of groups in the solute average out. The other three groups in **5.30** are very different in size and polarity and this further limits the relative orientations and maximises the diastereomeric differences for enantiomeric solute molecules in solvent-solute complexes. Finally the *ring current* of the aromatic ring is very effective in inducing differences in ^{1}H chemical shifts of groups in enantiomeric solutes. The commonest chiral shift reagents are paramagnetic chiral 1,3-diketone complexes of lanthanides such as europium or praseodymium, e.g., **5.31**. They are effective for molecules that can coordinate *rapidly* and *reversibly* with the lanthanide (increasing its coordination from 6 to 7), e.g., ethers, alcohols, ketones, esters, and amines. The use of chiral solvents and chiral shift reagents depend on mobile equilibria, very fast on the n.m.r. timescale, between free and complexed solute molecules.

There is a subtle but very important advantage in methods depending on chiral environments rather than diastereomeric derivatives to differentiate enantiomers. High enantiomeric excess in the chiral chromatographic stationary phase, solvent or shift reagent is *not* essential, in contrast to the use of chiral reagents to make diastereomeric derivatives. The *separation* of bands for enantiomers in chromatographic analyses, or *differences in chemical shifts* in n.m.r. spectra, depend on the enantiomeric excess for the chiral 'reagent' but the *analysis* depends on the areas of the bands or peaks.

Further reading

Chirality in Industry: the Commercial Manufacture and Applications of Optically Active compounds, ed. A.N. Collins, G.N. Sheldrake, and J. Crosby, Wiley, Chichester, U.K., 1992.

J. Jacques, A. Collet, and S.H Wilen, *Enantiomers, Racemates and Resolution,* Wiley, New York, 1981.

6 Stereospecific and stereoselective reactions

6.1 Some definitions

Description of the stereochemical outcome of reactions uses a number of terms and you may not be familiar with them all.

A substitution that replaces one ligand at a stereogenic sp^3 centre may proceed with *retention* (also referred to as *suprafacial* when part of a sigmatropic rearrangement: Section 6.5) or *inversion* of configuration (*antarafacial*). By this we mean that the arrangement of the other three ligands is (retention) or is not (inversion) left unchanged (Fig. 6.1). It does not refer to any *descriptor* that may apply to stereocentres in the molecule (see Problems at the end of this Chapter).

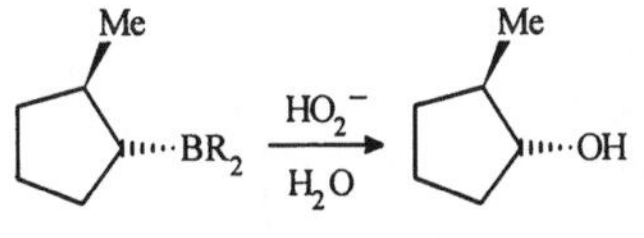

More formally: a reaction Cabcx to Cabcy is *homofacial* (= retention), or *heterofacial* (= inversion), with respect to the plane abc if x,y are on the same side, or on opposite sides, of abc.

Rearrangement from B to O, with retention of configuration (suprafacial) at C.

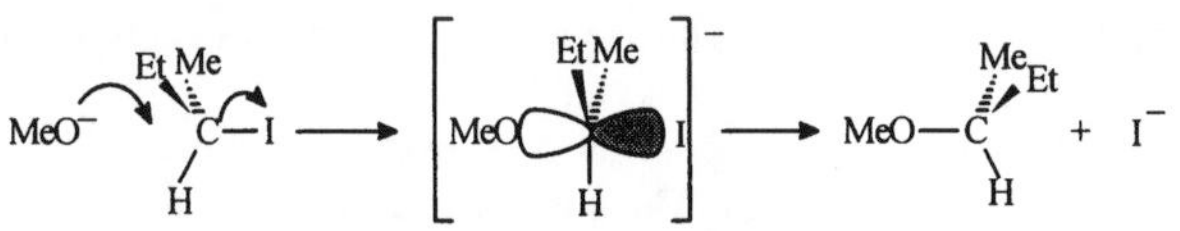

S_N2 with inversion of configuration (antarafacial) at C.

Fig. 6.1. Reactions with *retention* and with *inversion* of configuration at sp^3 C

In reactions at sp^2 C atoms retention is common but clear cut inversion is rare, hence the relatively obscure example (Fig. 6.2).

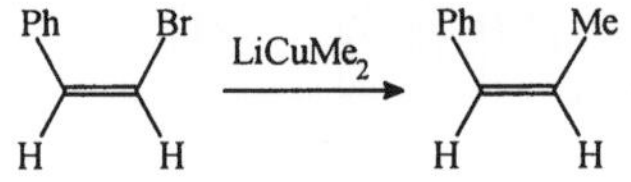

Retention at sp^2 C

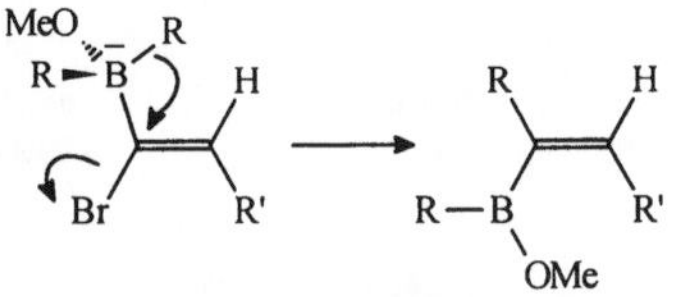

Inversion at sp^2 C

Fig. 6.2. Retention and inversion of configuration at sp^2 C

Reactions that take place at two sites, e.g., the ends of a π-system or of a σ-bond (Fig. 6.3) may be *suprafacial* or *antarafacial*, mainly in the context of pericyclic reactions (Section 6.5, pages 74–76). Additions to, and eliminations forming, one C=C are called *syn* (or *cis*) and *anti* (or *trans*) in other contexts (Fig. 6.3).

(a)   (b)   (c)

Fig. 6.3. (a): *Suprafacial (syn, cis)* and *antarafacial (anti, trans)* additions to a π-bond; *syn, cis* and *anti, trans* are also used for eliminations. (b), (c): *Suprafacial* and *antarafacial* opening of a σ-bond.

6.2   Stereospecificity and stereoselectivity

The terms stereospecific and stereoselective have been used in a number of ways: some chemists prefer not to use stereospecific at all. A common definition is that a reaction is stereospecific if two stereoisomeric reactants give distinct stereoisomeric products, as in Fig. 6.4. In contrast, a reaction is stereoselective if stereoisomeric products are formed in unequal amounts irrespective of any stereoisomerism in the reactant(s). A disadvantage of these definitions is that a reaction that is stereospecific for two stereoisomeric alkenes must be called stereo*selective* for an alkyne, or for a small ring alkene, which have no stereoisomers, even though the mechanism is the same.

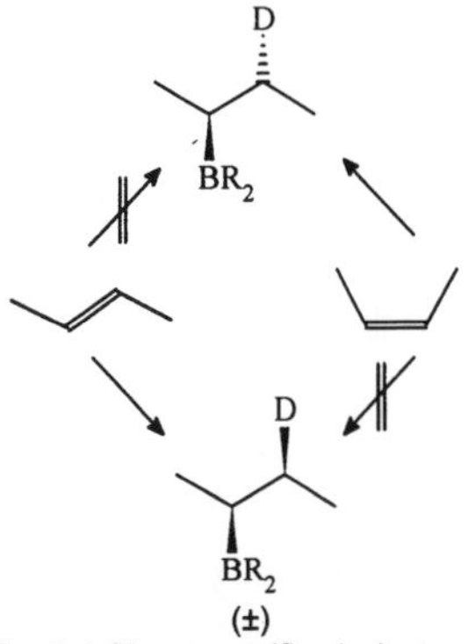

I will use stereospecific to describe reactions in which the stereochemical outcome is determined by *the orbitals of the functional group(s) necessarily involved in the reaction*, as in the examples given in Figs. 6.1–6.3. The addition of a borane to a π-bond (Figs. 6.3(a), 6.4) is a particularly clear example of the same stereochemical outcome for a reaction of alkenes and alkynes. In a stereoselective reaction possible stereoisomeric products are formed in unequal amounts due to the effects of parts of the reactants (or catalysts, solvents) *outside* the orbitals of the functional groups: see, e.g., Figs. 6.41, 6.42, page 79; Fig. 6.54, page 83; Fig. 6.58, page 84. *Multistep* reactions will be stereospecific overall only if each step is stereospecific, as in the electrophilic addition of Br_2 to C=C (Fig. 6.3(a), and Fig. 6.22, page 73). One advantage of these definitions is that they naturally allow many reactions to be described as both stereospecific *and* stereoselective when two or more stereogenic centres are formed (see, e.g., the Diels Alder reaction in Fig. 6.30).

Sections 6.3–6.7 describe the stereo-specificity and -selectivity of fairly simple reactions. Stereospecific reactions are very useful in synthesis if the stereochemical outcome is the one you want but if it is not there may be no way to get a desired product efficiently. In contrast, a nonstereospecific reaction with a zero, low, or qualitatively unwanted stereoselectivity when first tried, may have the potential to be developed to give high selectivity in the required sense. For this we need to discuss stereoisomeric transition states in some detail, in Section 6.8.

Fig. 6.4 Stereospecific *cis* deuteroboration of C=C by reaction with R_2BD

6.3   Polar reactions at carbon

Simple polar reactions may be classified as (i) substitutions, (ii) additions, and (iii) eliminations: in this Section the latter will be discussed together.

Substitutions in an organic molecule are classified according to the nature of the reagent (nucleophile or electrophile). Inversion in the S_N2 reaction is one of the most reliable stereochemical results in organic chemistry. The trigonal bipyramidal transition state requires an increase in the coordination of the central carbon atom and explains the great sensitivity to steric hindrance shown by S_N2 reactions (Fig. 6.5). Note that the large hindrance caused by the methyl groups in Me_3CCH_2Br does not force a change from inversion to retention of configuration.

S_N1 (S_N2): Substitution Nucleophilic uni- (bi-)molecular.
S_Ni: Substitution Nucleophilic Internal.
S_E2 Substitution Electrophilic bimolecular.

The S_N2 reaction has been shown to proceed with inversion in methyl and primary alkyl halides by using isotopic substitution to produce chiral molecules such as CHDTI and PhCHDCl.

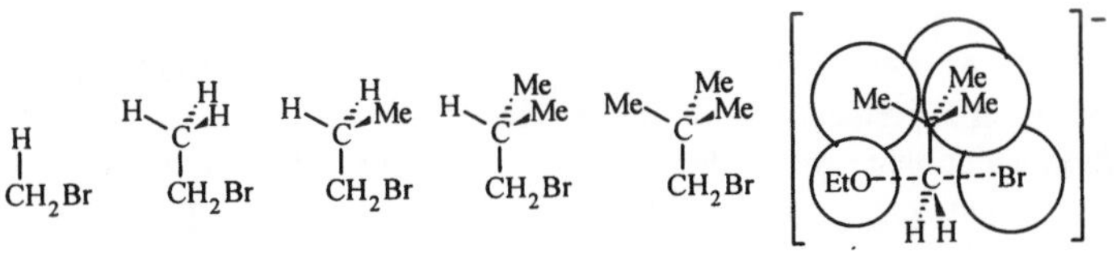

| 1 | 6×10^{-2} | 1.6×10^{-2} | 1.7×10^{-4} | 2.4×10^{-7} |

Fig. 6.5. Relative rates of S_N2 reaction of primary alkyl bromides with NaOEt/EtOH/56°C

S_N2 reactions of allyl halides can occur stereospecifically with inversion at C-1 or with varying stereoselectivity at C-3 (S_N2'). Much more widely useful substitutions in allyl compounds depend upon *catalysis* by Pd^0 complexes such as $Pd(PPh_3)_4$ (Fig. 6.6), for a wide range of nucleophiles and leaving groups. Such reactions are stereospecifically *suprafacial* overall with respect to the allyl system, a result of *two* stereospecific antarafacial steps. The second step is regioselective, mainly a result of steric effects.

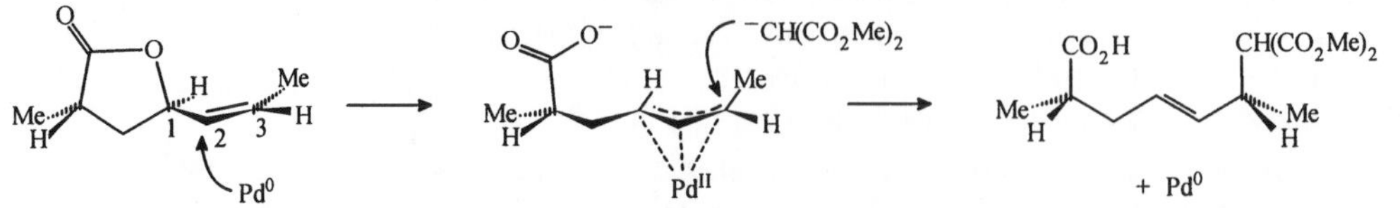

Fig. 6.6. Pd^0 catalysed allylic substitution with double bond shift: the other ligands on Pd, typically PPh_3, are omitted

Primary alkyl (other than methyl and ethyl) compounds probably always undergo S_N1 reactions with simultaneous rearrangement to a secondary or tertiary carbocation. Rearrangements are also common for secondary alkyl compounds: see L.M. Harwood's *Polar Rearrangements*, Oxford U.P., Oxford, 1992, for a detailed account.

S_N1 reactions of simple secondary and tertiary alkyl halides and sulfonates go via carbocations with low stereoselectivity favouring inversion. The latter is often attributed to a shielding effect of the departing anion but the true explanation is probably more complex for *s*-alkyl cations because many of them do not have simple structures (see Section 2.8, pages 12-13). S_N1 reactions tend to be *accelerated* by steric hindrance (relief of strain as the central C atom changes from sp^3 to sp^2).

A striking stereochemical feature of S_N1 reactions is the need for a planar central C^+ atom. Consequently there is a large increase in ring strain when many bridgehead halides ionise and their reactivity is then very low.

S_N2 reactions with inversion are impossible for the bridgehead bromides.

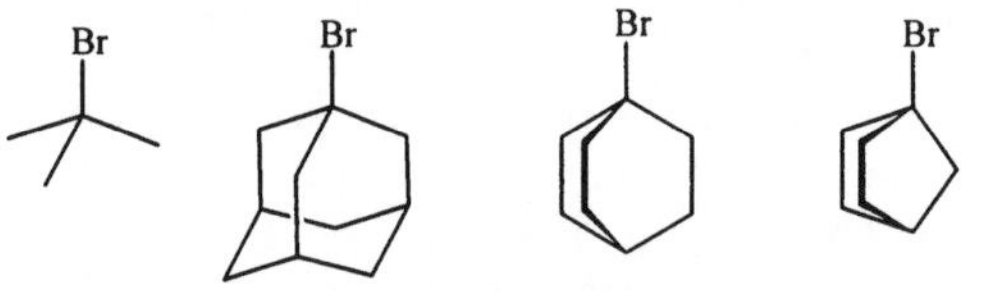

| 1 | 10^{-2} | 10^{-6} | 10^{-13} |

Fig. 6.7. Relative rates of solvolysis of t-butyl and bridgehead bromides in H_2O-EtOH

The S_Ni reaction is a special case of the S_N1 mechanism, with stereoselective *retention of configuration* (Fig. 6.8). In the absence of effective competition from the solvent or other nucleophiles the leaving group -OSOCl breaks down to generate Cl^- on the same side of C^+ as the original leaving group. If a weak base such as pyridine is present it converts HCl into Cl^- ions which give substitution with mainly *inversion*.

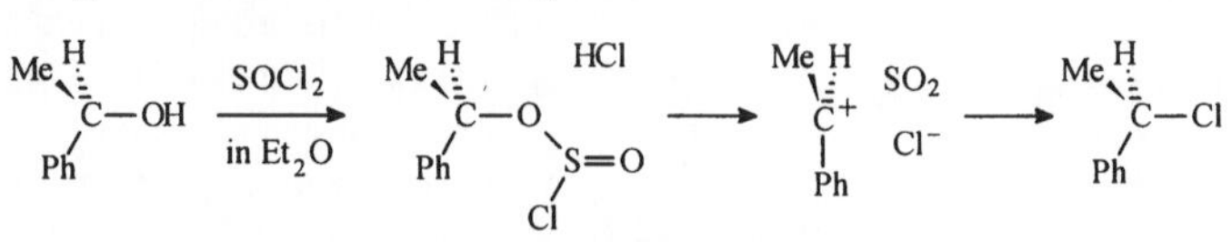

anti 7 *syn*

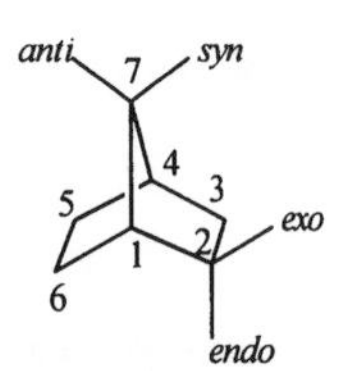

Orientations of substituents in bicyclic systems:
syn(*anti*) groups on *highest* numbered bridge towards (away from) *lowest* numbered bridge; *exo* (*endo*) groups on *lowest* numbered bridge towards (away from) *highest* numbered bridge.

Fig. 6.8. An S_Ni reaction

Some S_N1 reactions involve *intra*molecular nucleophiles (including a pair of bonding electrons) and the stereochemical outcome may then be very informative. A classic example is the solvolysis of homochiral *exo*-norbornyl (bicyclo[2.2.1]hept-2-yl) brosylate **6.1** (Fig. 6.9). The ionisation involves a stereospecific participation of the σ-electrons of the 1,6-bond to give an *achiral* cation that gives a 99.8% racemic acetate **6.2**. The *endo* stereoisomer reacts ~ 1800 times slower but also gives a mainly racemic *exo* product.

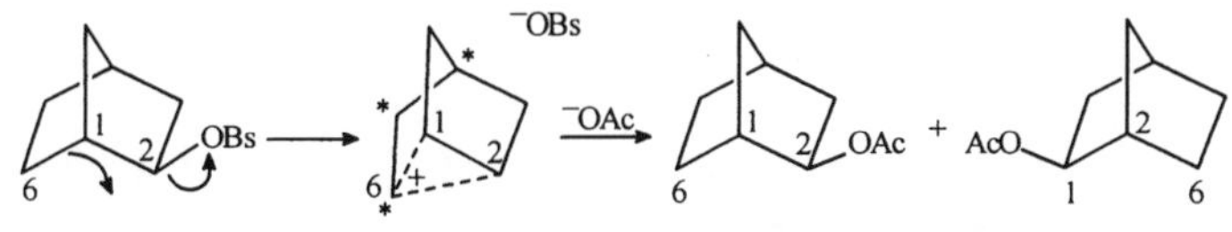

Fig. 6.9. The S_N1 reaction of *exo*-norbornyl brosylate in a CH_3CO_2H/CH_3CO_2Na buffer via an achiral cation, with a plane of symmetry through the starred atoms and mid-point of the 1,2-bond. [Bs = $-SO_2C_6H_4Br$ (p)]

Analogous stereospecific assistance to ionisation has been observed for π-(Fig. 6.10) and non-bonding electrons in solvolyses.

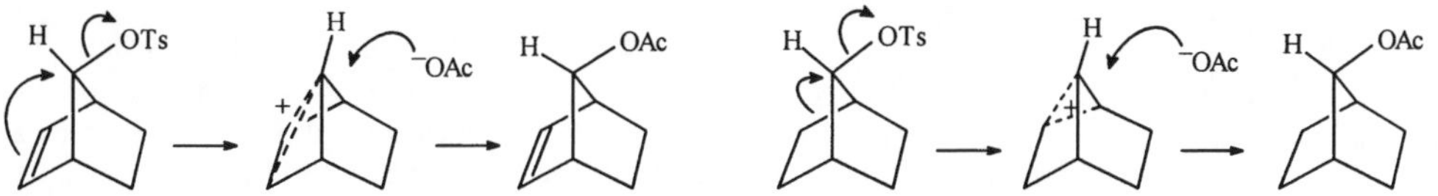

Relative rates: ~10^{10} 1

Fig. 6.10. Participation by π–electrons giving acceleration and overall retention. Retention in the saturated compound has been demonstrated by ^2H labelling and is probably caused by σ-participation. [Ts = $-SO_2C_6H_4Me$ (p)]

Electrophilic substitutions at sp^3 carbon atoms are varied in mechanism and stereochemistry. Some reactions are multistep through sp^2-carbon intermediates, e.g., the enolization and subsequent reactions of ketones with electrophiles, and are therefore not stereospecific. Single step S_E2 reactions at sp^3 usually go with *retention*. Quite apart from the exceptions this is not nearly as useful as inversion in S_N2 reactions because it is often difficult or impossible to get stereoisomerically pure reactants, particularly most RLi and RMgX compounds which are usually generated from the metals and alkyl halides via alkyl radicals (see Fig. 6.25, page 73, for an exception).

Polar substitution at unsaturated centres

Nucleophilic substitution in simple vinyl derivatives is unfavourable. S_N1 reactions with leaving groups such as $CF_3SO_3^-$ go via *linear* vinyl cations with low stereoselectivity (see also Section 2.8, page 13). S_E2 reactions go with retention.

Nucleophilic substitutions at unsaturated centres in conjugated systems with electron withdrawing groups, however, are favourable, usually bimolecular via association-dissociation, and can be stereoselective. When a nucleophile adds to one face of the π-system the *very short lived intermediate* is formed initially with the leaving group, e.g., Cl in Fig. 6.11, on the other face of the π-system. Stabilizing π→σ*-hyperconjugation between the newly electron rich π-system $^-O-C=C$ and the C-Cl bond will *increase* if the latter swings down ~60° (Fig. 6.11). If this is followed by rapid elimination of Cl⁻ then the overall substitution will be strongly stereoselective with retention (Fig. 6.12). A poorer leaving group than Cl may not leave *as rapidly* and then further bond rotation could occur allowing substitution with low stereoselectivity.

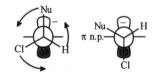

Fig. 6.11. After the addition of the nucleophile the C–Cl rotates to maximize π→σ* interaction with the now electron rich π-system $C=C-O^-$. (π n.p: nodal plane of π system)

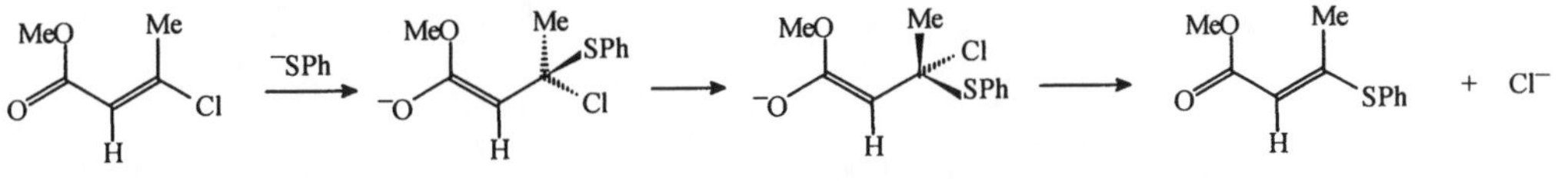

Fig. 6.12. Nucleophilic substitution with retention in a conjugated vinyl system

Polar eliminations and additions

1,2-Eliminations involve loss of substituents from neighbouring atoms to generate a double bond. The less common 1,1-eliminations (forming carbenes) or 1,3-eliminations (which may form carbenes, cyclopropanes, or 1,3-dipoles) have much less stereochemical interest.

1,2-Eliminations are also called α,β- or β-eliminations, or simply eliminations when it is unnecessary to distinguish them from 1,1- or 1,3-eliminations.

Eliminations are stereospecific if both substituents leave simultaneously (as in E2 and cyclic eliminations through five-membered ring transition states), otherwise they tend to have low stereoselectivity (as in the two step E1 reaction). Cyclic eliminations through six-membered ring transition states show varied stereoselectivity.

Polar 1,2-eliminations

E1: Elimination **unimolecular** (via ionisation to a carbocation in the rate determining step).
E2: Elimination **bimolecular**.

E2 reactions ideally require the torsion angle between the breaking bond vectors to be 180° or 0° (Fig. 6.13) in order to generate an untwisted double bond. Base promoted *anti* E2 reactions are usually quite strongly preferred to *syn* eliminations in acyclic and six-membered ring compounds.

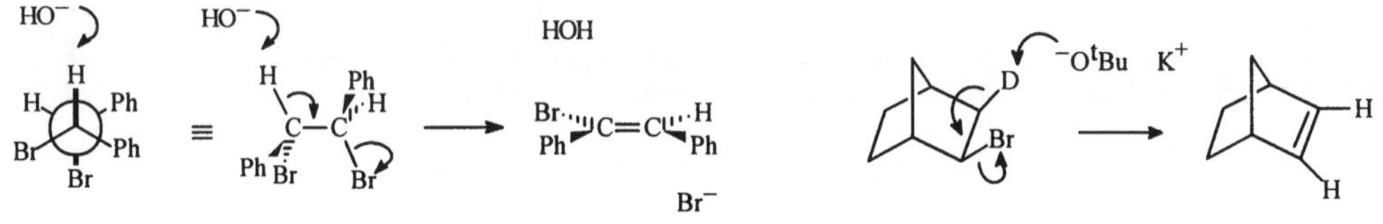

Fig. 6.13. *Antiperiplanar* and *synperiplanar* E2 reactions

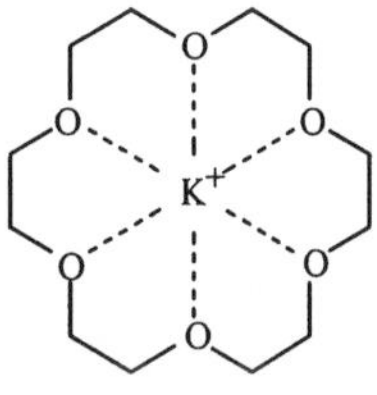

6.3

Syn (or *cis*) E2 reactions are found when (i) *anti*-elimination is impossible (see Problem 1(c), page 85), (ii) the groups to be eliminated are held *synperiplanar* (*syn* elimination at a torsion angle of 0° is better than *anti* elimination at 120°), as in Fig. 6.13 and in other five-membered ring compounds, or (iii) strong ion pairing occurs in relatively poor dissociating solvents (Fig. 6.14). The mixture of *syn* and *anti* E2 reactions was shown by D labelling. Sequestering the K^+ with a crown ether, as in **6.3**, cuts out the *syn* mechanism.

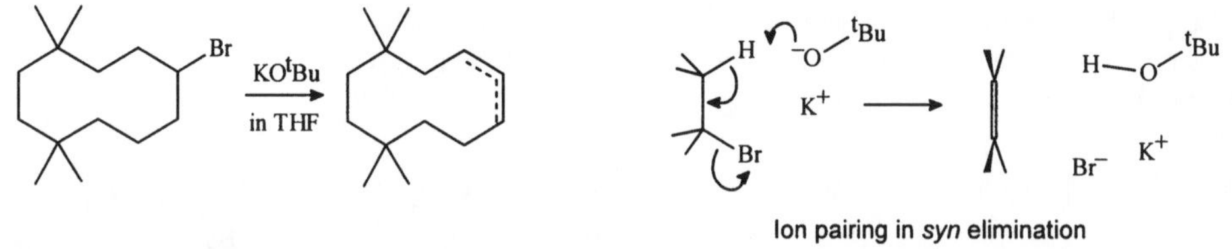

Ion pairing in *syn* elimination

Fig. 6.14. Both *syn* and *anti* eliminations, to give a mixture of alkenes, occur with KO^{t}Bu in tetrahydrofuran (THF)

Sayzeff's Rule: In an elimination the double bond goes preferentially to the most highly substituted carbon. When stereoisomeric alkenes result the more stable predominates.

In cyclic systems there can be a clash between Sayzeff's Rule for *regio*selectivity and the preference for *anti* elimination in E2 reactions. *Anti* elimination usually predominates (Fig. 6.15). E1 reactions via a carbocation commonly have low stereoselectivity, with the most stable regio- and stereoisomeric alkene predominating in most simple systems (Sayzeff's Rule).

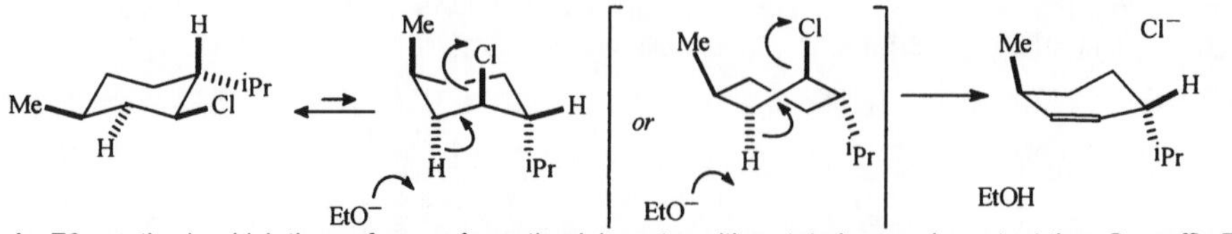

Fig. 6.15. An E2 reaction in which the preference for *antiperiplanar* transition state is more important than Sayzeff's Rule. It is uncertain whether the transition state has a very strained chair (three axial groups) or a twist conformation.

Reductive eliminations, e.g., formation of alkenes from 1,2-dibromides and similar compounds, follow two mechanistic pathways, one a stereospecific polar reaction and the other via one electron reductions, with low stereoselectivity. In the polar mechanism a *nucleophilic* reducing agent, e.g., I⁻, induces *anti* elimination (Fig. 6.16). *Anti* elimination is impossible in a *cis*-1,2-dibromide and it is believed that a slow S_N2 reaction (with inversion) forms a *trans*-bromoiodide which then undergoes reductive elimination.

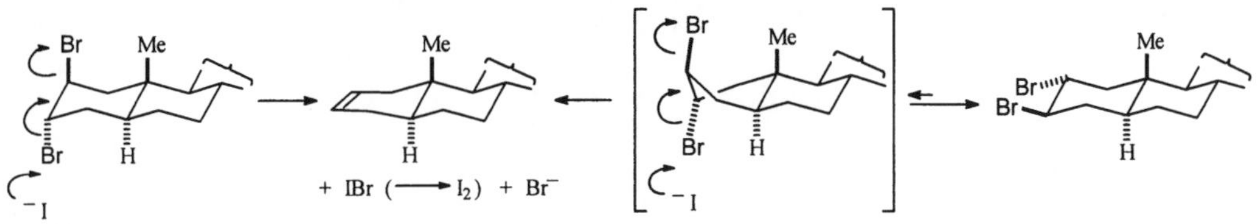

6.4: Relative rate: ~100   **6.5**: Relative rate: 1

Fig. 6.16. Debromination of diaxial (**6.4**) and diequatorial di-bromides (**6.6**). The relative rates are probably determined mainly by the unfavourable chair to twist change needed to achieve *antiperiplanar* arrangement of the two C–Br bonds in **6.5**. The reaction could involve I⁻ attacking either of the Br atoms. [N.B. The ring system is as in Fig. 6.21.]

Thermal unimolecular eliminations via five-membered ring transition states are either stereospecific pericyclic reactions (reverse of cycloadditions: see below) or show high stereoselectivity, because of the geometrical constraint when a five- or six-membered ring ester eliminates via a bicyclic transition state. The reaction conditions can be very mild, notably using a selenoxide (Fig. 6.17) generated *in situ* near room temperature. Thermal eliminations via four- or six-membered ring transition states may be stereospecifically *syn* pericyclic reactions or of moderate stereoselectivity.

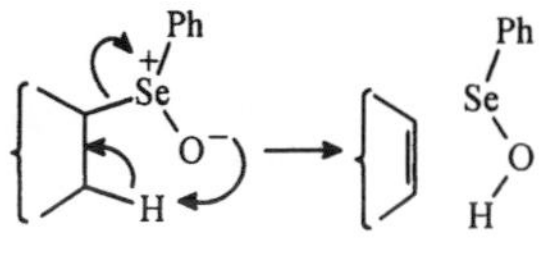

Fig. 6.17. A thermal *syn* elimination

Eliminations are not limited to the formation of a single double bond in a reaction. Stereospecific multiple eliminations, called *fragmentations*, can occur when several coplanar bonds are suitably aligned with leaving groups. A planar zigzag alignment of orbitals is the most favourable (Fig. 6.18).

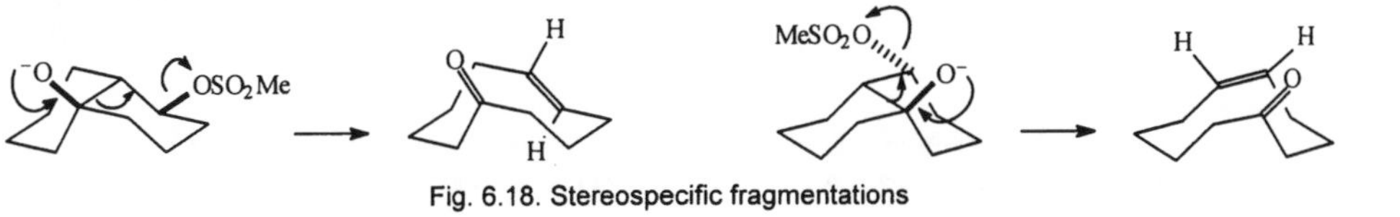

Fig. 6.18. Stereospecific fragmentations

Polar additions to alkenes

Polar additions of HX to alkenes take place in two steps, addition of a proton to give a carbocation which reacts with a (weak) nucleophile HX, e.g., H_2O, or X⁻, e.g., Cl⁻, with little stereochemical connection between them. The stereoselectivity can vary greatly with conditions (Fig. 6.19).

Compare the variable and often low stereoselectivity of these additions with the stereoselectivity of the reverse reaction, E1.

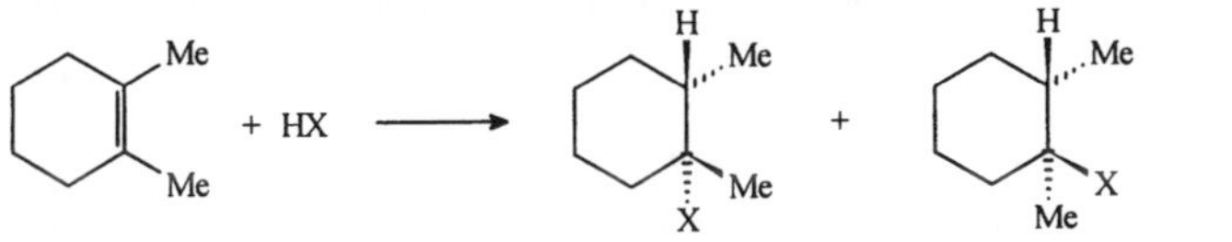

X: OH	H_2O + HNO_3
X: Cl	HCl in CH_2Cl_2 / – 98°C
X: Cl	HCl in Et_2O /0°C

~50%   ~50%
                        Main product

Main product

Fig. 6.19. Variable stereoselectivity in addition of acids to an alkene

Many electrophilic reagents, e.g., Br_2, add to alkenes via bridged cation intermediates. The latter usually react with nucleophiles *stereospecifically* to give *anti* addition and the diaxial product is formed *stereoselectivity* in six-membered rings, although the diequatorial is usually more stable (Fig. 6.20).

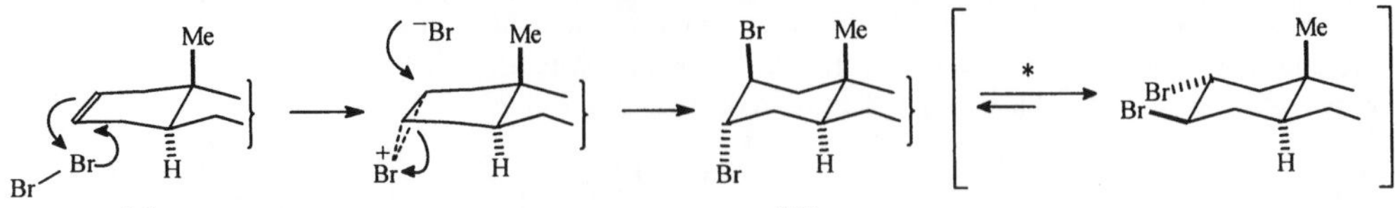

6.6   **6.7**

Fig. 6.20. *Trans*-diaxial addition of Br_2 (~90% **6.7**). The ring system is that in Fig. 6.21.
* This is a slow stereoisomerization, *not* a conformational change, that moves the substituents from axial to equatorial.

Overall *anti* addition is also observed when the three-membered ring compound is a stable intermediate, e.g., an epoxide (Fig. 6.21). The epoxide is formed stereo*selectively* on the less hindered face of the C=C by a stereo*specific syn* addition. The ring opening is favoured by a collinear Br–C–O unit in the transition state, as in an S_N2 reaction. This leads from the protonated epoxide to an *antiperiplanar* grouping Br–C–C–OH via chair- (preferred) or twist boat-like transition states to the primary products **6.8** (the main product) and **6.9**. The latter then changes from twist to chair to give **6.10**.

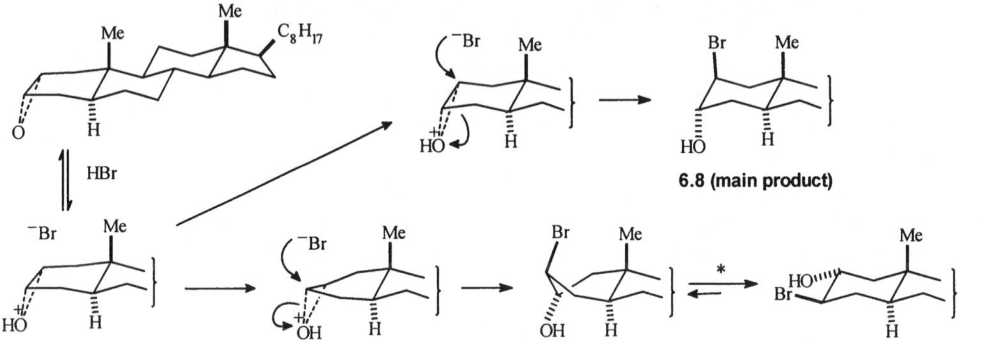

6.8 (main product)

6.9   **6.10 (minor product)**

Fig. 6.21. Stereospecificity and regioselectivity in the reaction of an epoxide with HBr.
* This is a rapid *conformational* change: the ring changes from twist to chair and the added substituents become equatorial.

See L.M. Harwood, *Polar Rearrangements*, Oxford University Press, Oxford, 1992.

1,5-Dienes and related polyenes can undergo spectacular multiple electrophilic additions (formally the reverse of fragmentations) to generate polycyclic compounds stereospecifically.

Addition of a carbene to an alkene is *either* stereospecifically *syn* in a single step if the carbene is a singlet *or* non-stereospecific in two steps with a low stereoselectivity if the carbene is a triplet, i.e., a diradical (Fig. 6.22). The same mixture of *cis*- and *trans*-1,2-dimethylcyclopropane is obtained from *cis*- or *trans*-2-butene.

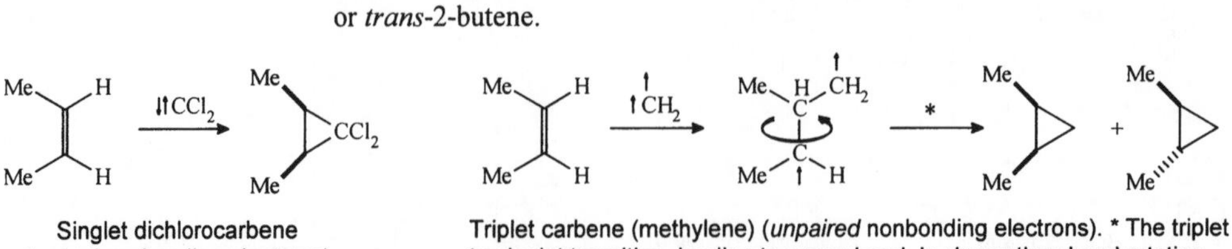

Singlet dichlorocarbene
(*paired* nonbonding electrons)

Triplet carbene (methylene) (*unpaired* nonbonding electrons). * The triplet to singlet transition, leading to a new bond, is slower than bond rotation.

Fig. 6.22. Additions of singlet and triplet carbenes to a simple alkene

6.4   Radical reactions

Simple alkyl radicals are planar or nearly so at the radical centre (see Section 2.8, page 13) and there is usually no relationship between the configurations of an sp^3 reaction site before and after a radical substitution. Relatively little strain results, however, when radicals are constrained to be pyramidal, e.g., at bridgeheads, in contrast to carbocations (see Fig. 6.7, page 68). The addition of HBr to C=C in diastereomeric alkenes at room temperature (Fig. 6.23) is

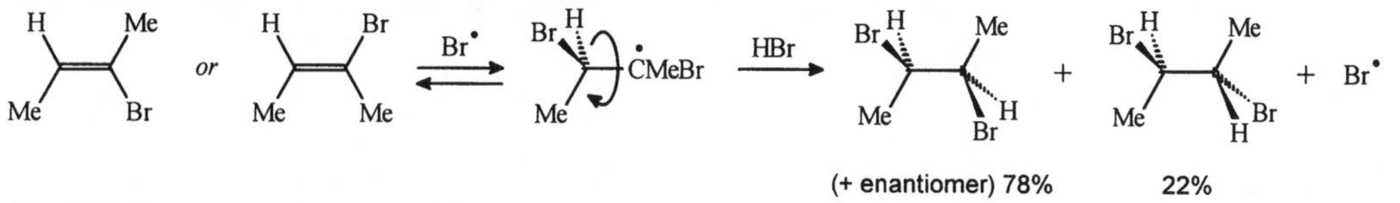

(+ enantiomer) 78%   22%

Fig. 6.23. Stereoselective addition of HBr to Z- and E-bromobut-2-ene at 25°C giving 78% of *racemic* 2,3–dibromobutane

typical of most intermolecular radical additions to alkenes. At –70°, however, the internal rotation is slow enough for addition of HBr to show marked stereospecificity (Fig. 6.24), possibly resulting from a weak bonding interaction between bromine and the radical centre.

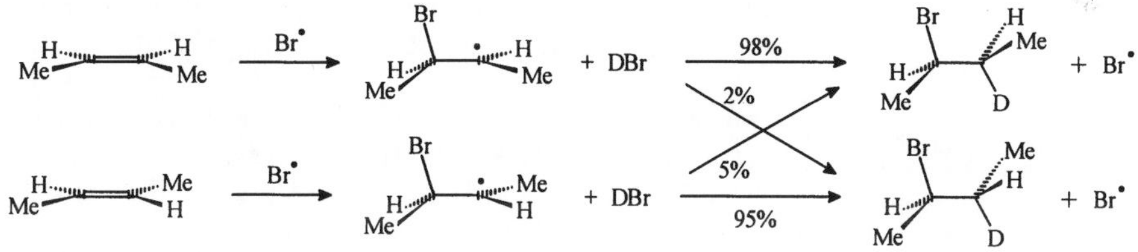

Fig. 6.24. Stereospecific addition of DBr to Z- and E-but-2-ene at –70°C (only one enantiomer of each product is shown)

Stereoselectivity in reactions via radicals

Although simple radicals are (almost) planar, cyclopropyl, α-oxy-radicals, and vinyl radicals tend to retain configuration and this leads to stereospecificity or stereoselectivity in many of their reactions at sufficiently low temperatures: see, e.g., Fig. 6.25.

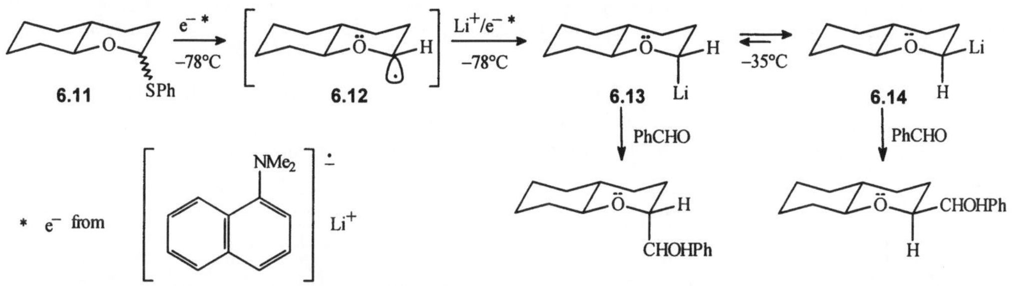

Fig. 6.25. The *pyramidal* α-oxyradical (**6.12**), stabilized by n→σ hyperconjugation, is formed stereo*selectively* from both diastereomers of the sulfide **6.11**. It reacts stereo*specifically*, with retention, with e⁻ and Li⁺ to give initially the *less* stable lithio-ether **6.13**. The latter rearranges at –35°C to **6.14**. **6.13** and **6.14** react stereo*specifically*, with retention, with, e.g., PhCHO.

Reactions of radicals can be very stereoselective when an additional fused ring is formed or when there is strong steric hindrance to one face of a planar radical centre. In Fig. 6.26 the radical cyclization gives a *cis* ring junction in **6.15** because it is difficult for the radical to reach round to the other side of the C=C. The formation of a five- rather than six-membered ring is usually preferred kinetically for such cyclizations (see Section 6.6, pages 76–77). The bridgehead Me and the new ring greatly hinder the *top* face of **6.15** and H is therefore added from Bu_3SnH to the other side to give the *trans*-decalin **6.16**. The latter can be converted into **6.17**, with a hindered *axial*–CH_2OH, that would be very difficult to prepare by non-radical reactions.

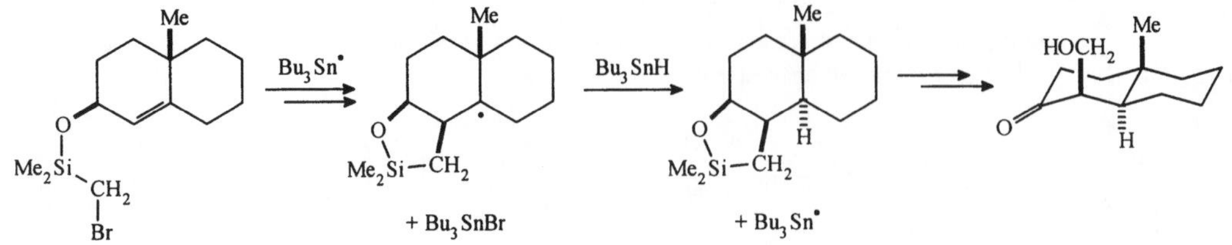

6.15 **6.16** **6.17**

Fig. 6.26. Examples of stereoselective radical reactions

6.5 Pericyclic reactions

Pericyclic reactions include several different classes of reactions, of which *cycloadditions*, *electrocyclic* reactions, and *sigmatropic* rearrangements are the most important, that exhibit striking stereo-specificity and -selectivity. All thermal pericyclic reactions may be regarded as having *aromatic* transition states formed from a suitable number of electrons in a closed conjugated cycle of orbitals. The cycle of orbitals can have one or the other of two topologies, Hückel or Möbius. In the Hückel transition states the basis set of atomic orbitals can be chosen to have no phase dislocations (just as in a stable Hückel aromatic system such as benzene), e.g., in the *disrotatory* cyclization of octa-2,4,6-triene (Fig. 6.27). The Hückel transition state is aromatic if it has $4n + 2$ electrons (σ as well as π; n is an integer).

The orbitals shown in the transition states in Fig. 6.27 to 6.31 and 6.33 are a basis set of AOs from which the MOs are derived.

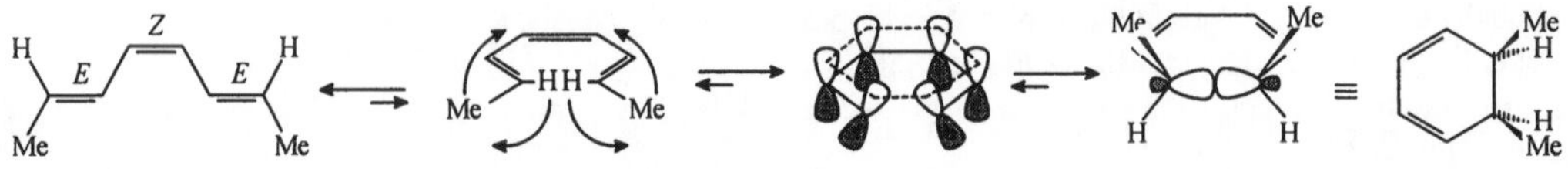

Fig. 6.27. *Disrotatory* ring closure of a 1,3,5-triene to a cyclohexa-1,3-diene via a Hückel transition state (for clarity the substituents are omitted from the transition state, here and in Figs. 6.28 to 6.31 and 6.33, to show how the basis set of atomic orbitals overlap).

In the Möbius transition state there must be a phase dislocation in the basis set of atomic orbitals, e.g., in the *conrotatory* ring opening of cyclobutene (Fig. 6.28) and the favoured number of electrons is $4n$.

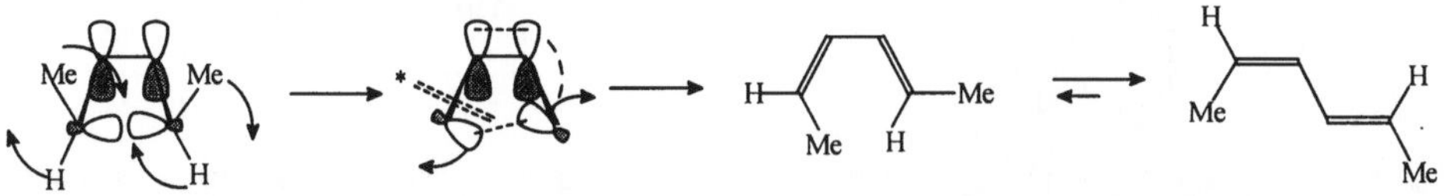

Fig. 6.28. *Conrotatory* ring opening of a cyclobutene to a buta-1,3-diene via a Möbius transition state. (* This marks the phase dislocation that characterises a Möbius transition state.)

Most cycloadditions go via Hückel transition states with 6-electrons. By far the most important is the Diels Alder reaction (Fig. 6.29). The Hückel topology for the transition state with six electrons leads to a stereospecific reaction that is suprafacial for both the ene and the diene.

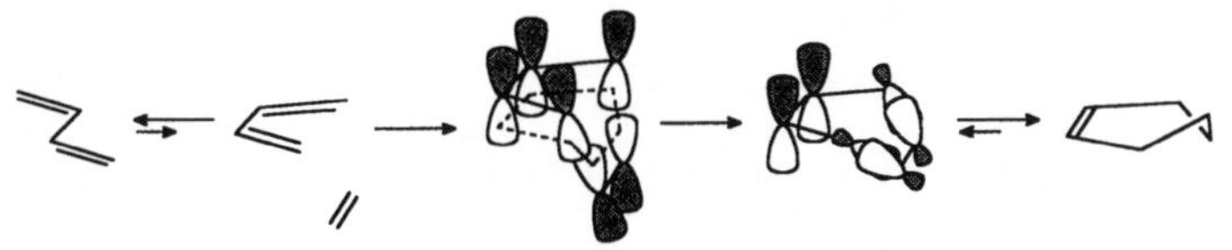

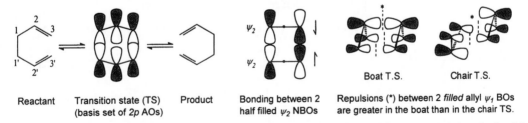

Fig. 6.29. The simplest Diels Alder reaction: cyclohexene is formed initially in a half boat conformation although the half-chair is more stable

Diels Alder reactions may also show *endo* stereoselectivity, which can be explained if we look outside the basis set of atomic orbitals. It results when suitable substituents are added to the basic system, as in Fig. 6.30, and is best understood using Frontier Molecular Orbital (FMO) theory. Ethene is a poor *dieneophile* in the Diels Alder reaction but its reactivity is greatly increased by π-electron withdrawing groups (π-EWG) which lower the energy of its LUMO, as in maleic anhydride (**6.18**). Cyclopentadiene is more reactive than buta-1,3-diene because the diene is held *s–cis* and its HOMO has a higher energy. The HOMO-LUMO interaction in the transition state is most important where the new bonds are forming but there is also weaker ('*secondary*') bonding between other pairs of atoms (marked *): this is not possible in the *exo* transition state.

See, e.g., I. Fleming, *Frontier Orbitals and Organic Chemical Reactions*, Wiley, Chichester, 1976.

s-cis: see Fig. 2.7, page 7.

HOMO: Highest Occupied MO.
LUMO: Lowest Unoccupied MO.
Reactivity in a Diels Alder reaction is inversely related to the *smaller* HOMO-LUMO energy gap for the two reactants.

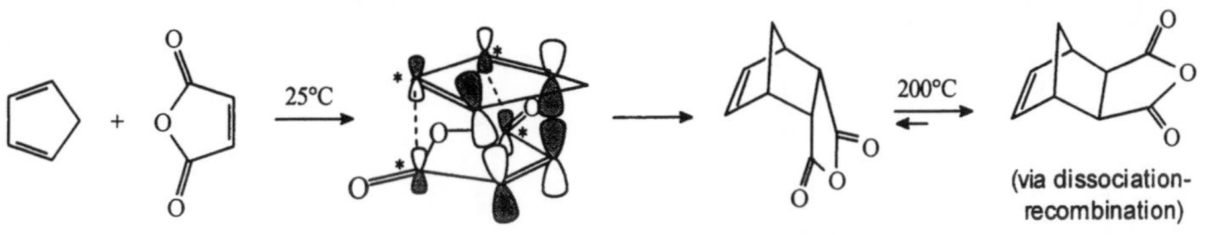

6.18   *endo* T.S   *endo*-**6.19**   *exo*-**6.19**

(via dissociation-recombination)

Fig. 6.30. *Endo* stereoselectivity in a Diels Alder reaction with a dienophile with a π-EWG: 'secondary orbital overlap' is believed to stabilize the *endo* transition state, although the product *endo*-**6.19** is less stable than *exo*-**6.19**.

In sigmatropic reactions either a σ bond moves to the other ends of two π systems or an atom or group moves to the end of a single π system. The [3.3]sigmatropic rearrangement, with one σ bond and two π bonds, is the most common type (Fig. 6.31), with the σ bond moving from C-1 to C-3 (hence [3.3]) in each half of the reactant. The transition state must have a Hückel topology to hold six electrons and can be looked at as being formed by

| Reactant | Transition state (TS) (basis set of $2p$ AOs) | Product | Bonding between 2 half filled ψ_2 NBOs | Repulsions (*) between 2 *filled* allyl ψ_1 BOs are greater in the boat than in the chair TS. |

Boat T.S.   Chair T.S.

Fig. 6.31. The simplest [3.3]sigmatropic rearrangement, with the transition state made from two allyl radicals (see Section 2.7, page 12), showing the bonding and anti-bonding interactions between the two allyl systems.

bringing two allyl radicals together. The principle *bonding* interaction is between the two half filled ψ_2 (non-bonding) MOs reacting with stereospecific suprafacial-suprafacial geometry. The latter can be achieved in two ways, giving a transition state with either a chair- or a boat-like shape. In the boat electrons in the *filled* ψ_1 allyl orbitals (BOs) repel one another and the chair transition state is more stable.

The preference for a chair transition state explains the stereoselectivity of such rearrangements. For example, the diene **6.20** gives >99% *Z,E*-**6.21** via a chair (Fig. 6.32) and <1% of the *E,E*- or *Z,Z*-**6.21** via boat transition states.

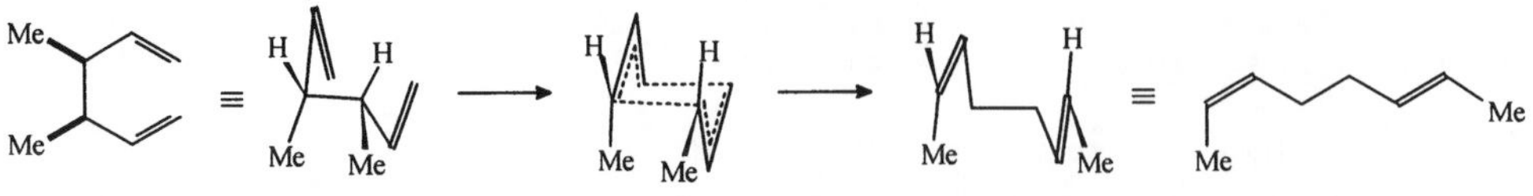

6.20   *Z,E*-**6.21**

Fig. 6.32. Stereoselectivity in the [3,3]-sigmatropic rearrangement of *meso*-3,4-dimethylhexa-1,5-diene

Fig. 6.33 shows how a sigmatropic rearrangement was used to correlate the absolute configurations of a chiral centre and a chiral axis.

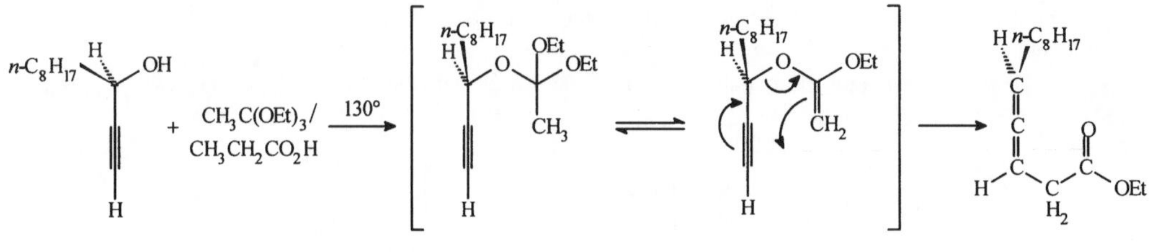

R   *M*

Fig. 6.33. An example of chemical correlation of configurations of a chiral centre and a chiral axis

6.6 Baldwin's Guidelines for ring closure

J.E. Baldwin, *J.C.S. Chem. Comm.*, 1976, 734, 736.

N.B. These reactions are all ring *forming* reactions except for *n*-endo-tet in which an acyclic reactant passes through a cyclic transition state to an acyclic product.

Reactions that are generally useful for intermolecular reactions sometimes fail, or are very slow, or take an unexpected course, in *intramolecular* reactions with to 3- to 7-membered ring transition states. These observations have been rationalized by a rather simple set of three guidelines. The reaction must be classified by three parameters: the ring size, *n*, the hybridization of the atom attacked by the reactive centre forming a new bond (using *tet* for sp^3; *trig* for sp^2; *dig* for *sp* hybridization), and the position of the *breaking* bond relative to the smallest ring formed: *exo* if that bond is exocyclic to the smallest ring formed, otherwise *endo*. The three guidelines are:

Guideline	System	Favoured	Disfavoured
1	*tet*	3- to 7-*exo*	5- to 6-*endo* (3-, 4-*endo*: unknown)
2	*trig*	3- to 7-*exo* 6- to 7-*endo*	3- to 5-*endo*
3	*dig*	5- to 7-*exo* 3- to 7-*endo*	3- to 4-*exo-dig*

The guidelines are rationalized by assuming that the bond forming sites approach *tet*, *trig*, and *dig* sites from quite strongly preferred directions (see Section 6.7 for additions to C=O). The best experimental evidence for the guidelines is for *nucleophilic* (Fig. 6.34) and *radical* centres: electrophilic reactant centres are much less constrained.

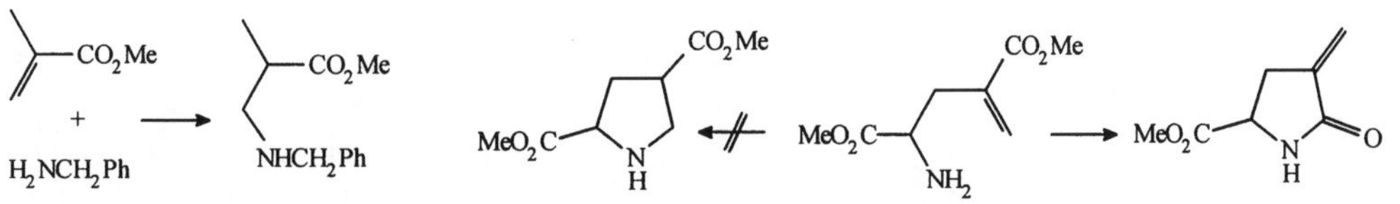

Acyclic: 1,4-addition to C=C–CO₂Me.

Cyclic: 5-exo-trig (on C=O) preferred to 5-endo-trig.

Fig. 6.34. Avoidance of a disfavoured 5-endo-trig cyclization

6.7   Reactions at C=O

Polar additions to C=C and C≡C are classified by the nature of the first species attacking the alkene because all such reactions begin with attack on a π-bond. In contrast, polar additions to carbonyl compounds are almost always classified as *nucleophilic additions*. This is obvious when addition begins with nucleophilic attack on the C=O, e.g., by ⁻C≡N, but not so clear when the electrophile initiates the reaction. When simple carbonyl compounds react with electrophiles the latter attack non-bonding electrons on O and do *not* break the C–O double bond (contrast with C=C, above). Such reactions with electrophiles, usually acids, are mainly important in speeding up subsequent nucleophilic attack at the C atom, breaking the C–O double bond and giving overall *(general* or *specific) acid catalysed nucleophilic addition* (Fig. 6.35).

Additions to C=O are *never* stereospecific because there is no observable difference between *syn* and *anti* addition. Many, however, are stereoselective.

The *trajectories* for simple nucleophiles, e.g., H⁻, approaching a C=O, as in H₂C=O (formaldehyde or methanal) have been calculated by quantum mechanics. Similar results have been inferred for C=O and amine nucleophiles from a statistical analysis of X-ray crystal structures of aminoketones. The key result is that at the nucleophile approaches the C=O bond vector at an angle close to 110°, maintaining this angle with the C–O bond to the end of the addition, and *not* along the axis of the C $2p_z$ orbital of the π-bond.

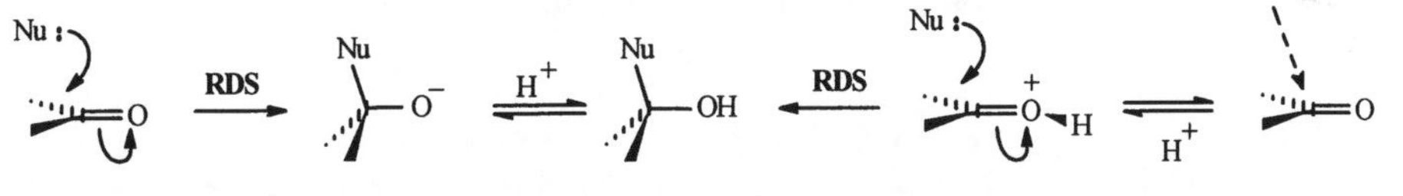

Fig. 6.35. Polar additions to C=O break the double bond only when the *nucleophile* (Nu) adds. Electrophiles, e.g., H⁺, add to an unshared pair of electrons on O *without* breaking the C=O. The dashed arrow shows the trajectory for nucleophilic attack.

An important class of carbonyl reactions involve one electron reductions. These reactions show stereoselectivity that can be strikingly different from nucleophilic addition and usually give the more stable of two diastereomeric alcohols (Fig. 6.36). In contrast the bulky hydride donor HBEt₃⁻ attacks from the less hindered face of the C=O to give the *less* stable alcohol.

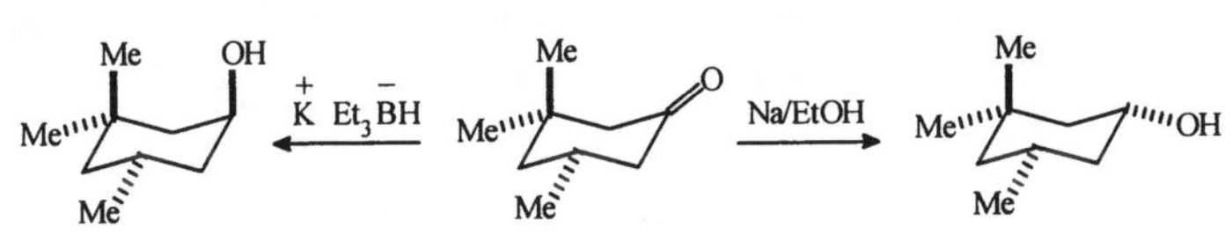

Fig. 6.36.

The following Section considers stereoselectivity, particularly in additions to C=O.

6.8   Stereoisomeric transition states

Enantiomeric transition states, like enantiomeric groups of molecules, have identical scalar properties, e.g., interatomic distances or energies.

Diastereomeric molecules or groups of molecules have the same constitutions for the corresponding molecules but diastereomeric transition states are less simple. Distinction between diastereomerism and constitutional isomerism in *molecules* depends upon assigning definite *bond orders* to bonds between pairs of atoms (Section 4.1, page 37). In diastereomeric transition states we must treat the *partial bonds* as equal and we often ignore some weak bonding interactions (Fig. 6.37: see also Fig. 6.30, page 75).

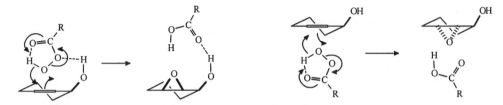

Fig. 6.37. Epoxidation of an allylic alcohol via *diastereomeric* transition states (the intermolecular hydrogen bond is favourable). The difference in hydrogen bonding is ignored in this definition.

Enantioselective reactions with low selectivity can be achieved using chiral solvents or circularly polarized (chiral) light.

Stereoisomeric transition states result from various combinations of (*pro*)*stereoisomeric* reactants, reagents, and catalysts. Reagents may be, e.g., (i) achiral, e.g., $NaBH_4$ (Fig. 6.39), (ii) chiral but with an inherently *achiral* species such as H^- (**6.22**, in Fig. 6.41) being transferred in the reaction, and (iii) chiral, with a chiral species being transferred to the reactant (Fig. 6.56, page 83).

In this Section I use standard molar free energy diagrams to illustrate different possibilities for stereoisomeric transition states. The difference in activation free energies, $\delta\Delta G^{\ddagger} = -RT\ln(k/k')$, is a measure of the ratio of rates constants (k/k') for the competing reactions.

Chiral centre created from an achiral starting molecule and achiral reagents, catalysts, etc.: generation of racemic products

If all reagents, catalysts and solvents are achiral then a prochiral substrate, in a reaction creating a chiral centre, will give *enantiomeric product* molecules P and P', via enantiomeric transition states with equal values of $\Delta G^{\ddagger}$ (Fig. 6.38), in *statistically equal* numbers. The fractional difference between the numbers of enantiomeric product molecules is immeasurably small for a typical laboratory scale reaction and will be ignored from now on. An example is the reduction of butan-2-one by $NaBH_4$ to *racemic* butan-2-ol (Fig. 6.39).

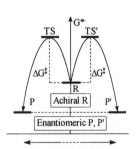

Fig. 6.38. Enantiomeric products P and P' from an achiral reactant R and achiral reagents, catalysts, etc., via enantiomeric transition states.

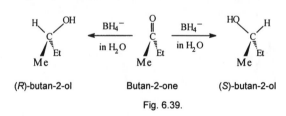

(*R*)-butan-2-ol   Butan-2-one   (*S*)-butan-2-ol

Fig. 6.39.

Reactions generating a chiral centre from an achiral substrate and chiral reagents or catalysts: enantioselective (or asymmetric) synthesis

If a reagent or catalyst is homochiral in a reaction generating a chiral centre in an achiral reactant the enantiomeric products will be formed in unequal amounts via *diastereomeric* transition states that *differ in energy* by $\delta\Delta G^{\ddagger}$ (Fig. 6.40). Such reactions are *enantioselective*. The *magnitude* of $\delta\Delta G^{\ddagger}$ is rarely predictable and even the *sign* is usually known only empirically.

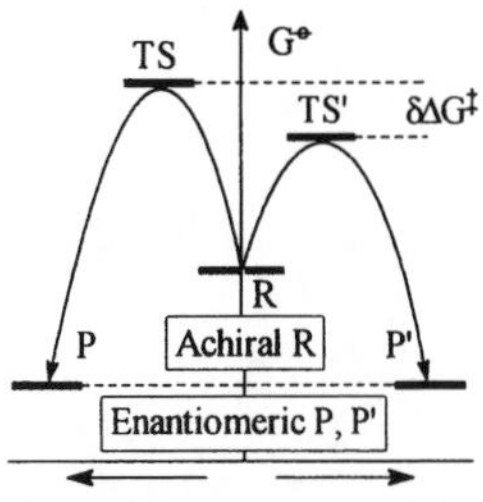

The following example shows how an achiral source of 'H^{-}' such as LiAlH$_4$ may be converted into a chiral reagent **6.22** by reaction with a chiral diphenol and ethanol (Fig. 6.41).

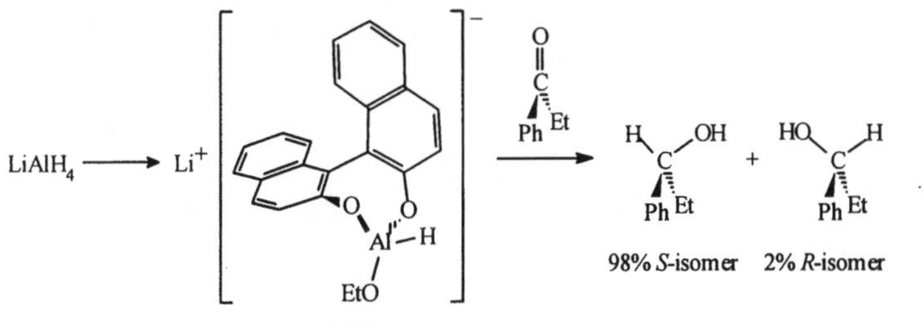

Fig. 6.40. As in Fig. 6.38 except that a chiral reagent or catalyst is used

6.22

Fig. 6.41. Use of a chiral reagent to transfer an achiral unit ('H^{-}') enantioselectively to a prochiral carbonyl group

N.B. The enantiomeric products P and P' (Figs. 6.40 and 6.41) are *not* exactly equal in energy in the chiral environment created by the reagent or catalyst but the effect is insignificant and will be ignored in what follows.

Effective homochiral reagents are usually expensive and chiral catalysts have much greater promise. Chiral homogeneous hydrogenation ('Wilkinson') catalysts are among the most effective for suitable substrates (Fig. 6.42).

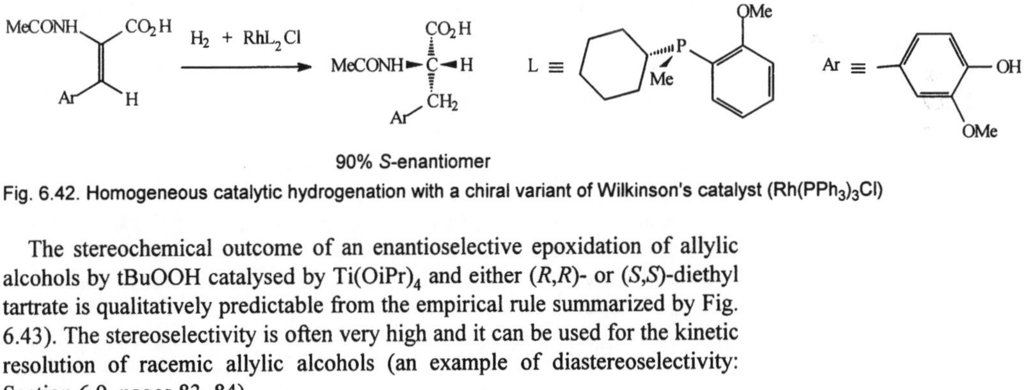

Fig. 6.42. Homogeneous catalytic hydrogenation with a chiral variant of Wilkinson's catalyst (Rh(PPh$_3$)$_3$Cl)

The stereochemical outcome of an enantioselective epoxidation of allylic alcohols by tBuOOH catalysed by Ti(OiPr)$_4$ and either (R,R)- or (S,S)-diethyl tartrate is qualitatively predictable from the empirical rule summarized by Fig. 6.43). The stereoselectivity is often very high and it can be used for the kinetic resolution of racemic allylic alcohols (an example of diastereoselectivity: Section 6.9, pages 83–84).

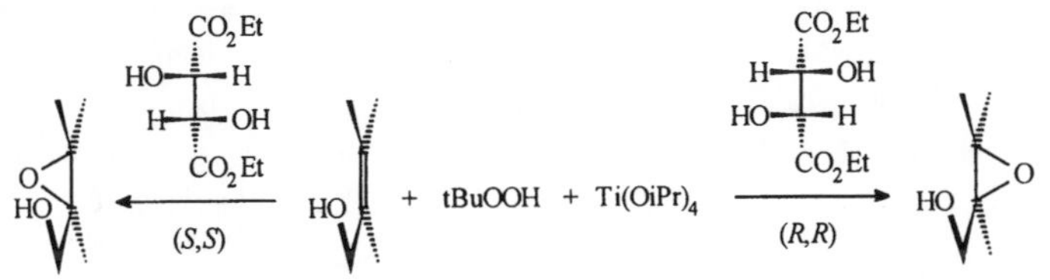

Fig. 6.43. Enantioselectivity in a epoxidation of an allylic alcohol (Sharpless's rule)

Enzymes, either isolated or in organisms, are widely applicable and can be spectacularly selective (see Fig. 5.12, page 65). Hydrolytic enzymes that accept a wide variety of esters or amides as substrates are more useful (see

Fig. 6.47). Yeasts are widely used as a chiral reducing agents (see Fig. 6.49).

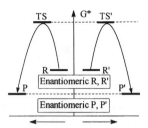

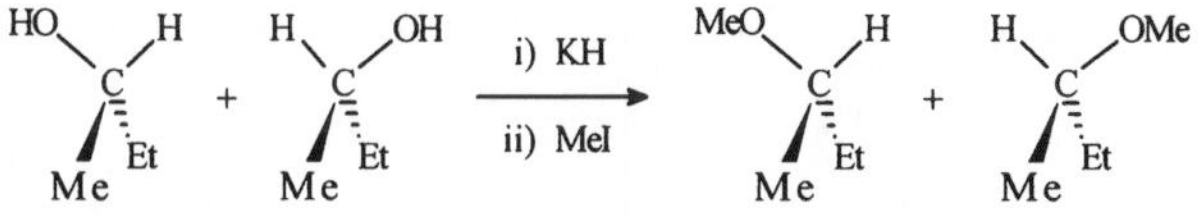

Fig. 6.44. Reaction of *enantiomeric* reactants R, R' with achiral reagents to give statistically *equal* amounts of *enantiomeric* products P and P'

Reactions of racemic compounds with achiral reagents, catalysts, and solvents

There are several types of transition states to consider when racemic compounds are reactants.

1. If the reagents, solvents, etc., are achiral then enantiomeric reactant molecules will form enantiomeric products via *enantiomeric transition states* at *equal rates*: $\delta\Delta G^{\ddagger} \equiv 0$ (Fig. 6.44). The apparently dull but important result is that a racemic reactant and achiral reagents give a racemic product (Fig. 6.45). This principle has been overlooked many times. As recently as 1994 an international journal published a paper claiming that enantioselective reactions could be induced by a magnetic field, which is *achiral*.

Racemic Racemic

Fig. 6.45. Reaction of a racemic reactant (butan-2-ol) with achiral reagents (KH, MeI) to give a racemic product (2-methoxybutane)

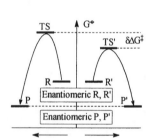

Fig. 6.46. Diastereomeric transition states converting *enantiomeric* reactants into enantiomeric products *enantioselectively*

Reactions of racemic compounds with homochiral reagents, etc.: Kinetic resolution

If a reaction does *not* generate a new chiral centre but a reagent or catalyst is homochiral then enantiomeric substrates R, R' will give enantiomeric products P, P' via *diastereomeric transition states* (i.e., with *unequal* values of $\Delta G^{\ddagger}$) at *unequal* rates. Consequently, if such a reaction is stopped part way through the product will contain more of one enantiomeric form (P' in Fig. 6.46) than the other: this is *kinetic resolution*. This assumes that any group added to the reactant is inherently achiral.

There is a subtle but vital difference between Fig. 6.46 (*racemic* reactant) and Fig. 6.40 (a single achiral reactant). In the latter the rates of reaction are in a constant ratio (determined by $\delta\Delta G^{\ddagger}$ alone) throughout the reaction. In Fig. 6.46 the *relative* rate of formation of P' *falls* during the reaction because R' is used up faster than R and the relative rates depend on [R] and [R'], as well as $\delta\Delta G^{\ddagger}$. The selectivity is greatest at the beginning of the reaction, when [R] = [R'], but decreases thereafter. If the reactions are taken to completion P and P' are necessarily formed in *equal* amounts! In contrast the ratio [R]/[R'] begins at unity and *rises throughout the reaction*, without limit, although [R] falls. This type of reaction is much more effective for preparing an enriched sample of the slower reacting *reactant* than of the faster formed *product*.

The range of reagents and catalysts that give useful differences in rates for enantiomers is limited but many enzymes promote simple reactions such as hydrolysis of esters and amides (Fig. 6.47) with high enantioselectivity.

L- and *D*-*N*-acetylamino acid *L*-α-amino acid *D*-*N*-acetylamino acid *D*-α-amino acid

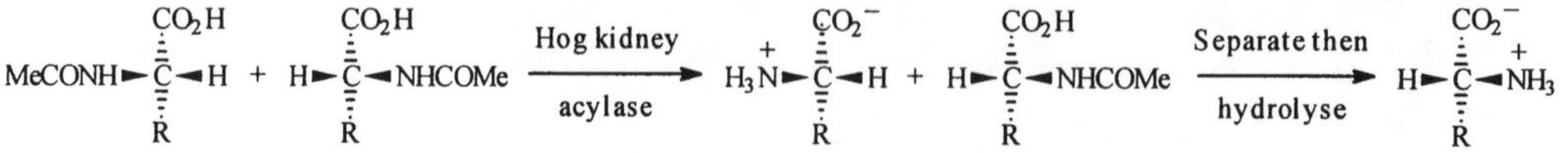

Fig. 6.47. Enantioselective hydrolysis by an enzyme. After separation of the *L*-amino acid and the *D*-*N*-acetylamino acid the latter can be hydrolysed *non*-enzymatically.

Kinetic resolution places higher demands on the magnitude of the differences in rates of reactions than does an enantioselective reaction of a single achiral reactant (Fig. 6.40).

A special case arises when the reactant racemizes readily, either spontaneously under the reaction conditions or with deliberate catalysis (Fig. 6.48). The enantiomeric reactants remain at equal concentrations and the ratio of rates of formation of enantiomeric products depends only on $\Delta G^{\ddagger}$ because $[R] \approx [R']$ throughout the reaction. The system resembles that in Fig. 6.40 and the enantioselectivity is *higher* than if R and R' do not interconvert (as in Fig. 6.46). A practical example is the reduction of a racemic β-ketoester (the enantiomers readily interconvert via enolization) by yeast (Fig. 6.49). This reaction exhibits enantioselectivity *and* diastereoselectivity (next sub-section). Homochiral hydrogenation catalysts can bring about analogous enantio- and diastereo-selective reactions.

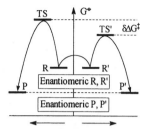

Fig. 6.48. Enantioselectivity when a readily racemized reactant reacts with a chiral reagent or catalyst

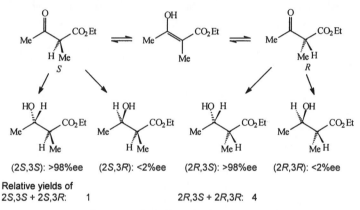

(2S,3S): >98%ee (2S,3R): <2%ee (2R,3S): >98%ee (2R,3R): <2%ee

Relative yields of
2S,3S + 2S,3R: 1   2R,3S + 2R,3R: 4

Fig. 6.49. Reduction of ethyl 2-methyl-3-oxobutanoate by yeast

Creation of additional stereogenic centres ('diastereoselectivity')

When a reaction creates one or more additional stereogenic centres in a chiral reactant R, diastereomeric products (P and P') are formed via diastereomeric transition states at *unequal rates*, i.e., in unequal amounts. The magnitude of $\delta\Delta G^{\ddagger}$ varies from the scarcely perceptible to as much as 25 kJ mol^{-1} in extreme cases, the latter giving high diastereoselectivity (~10^4:1 at room temperature). High selectivity is common in enzyme catalysed reactions but a rapidly growing number of non-enzymatic reactions are now known with rate ratios >10 for diastereomeric chiral products. In Fig. 6.50 the less stable product P' is formed preferentially through the lower energy transition state but there is no necessary relation between stabilities of the transition state and of the products. Understanding the origins of differences in energies of diastereomeric transition states is a major target for organic and biochemical research.

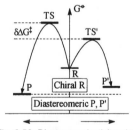

Fig. 6.50. Diastereoselectivity when reactant R undergoes competitive reactions forming diastereomeric products P and P': $\delta\Delta G^{\ddagger}$ is non-zero for *all* reagents

In some relatively simple reactions of this type there are useful empirical rules for predicting the *qualitative* selectivity.

The stereochemical course of nucleophilic addition to the ketonic C=O of an α-ketoester of a chiral alcohol with the configuration **6.23**, with groups with the relative sizes L>M>S (usually S=H), can be predicted assuming that the reactive conformation is **6.24** (Fig. 6.51: Prelog's Rule, using PhMgBr as a nucleophile). Such reactions have been used in two ways. The absolute configuration of the chiral C atom in the parent alcohol can be derived from the *known* configuration of the more abundant α-hydroxycarboxylic acid, obtained after hydrolysis of the ester. Alternatively, the reaction can be used as an enantioselective synthesis of α-hydroxy acids using an alcohol that gives rise to high diastereoselectivity.

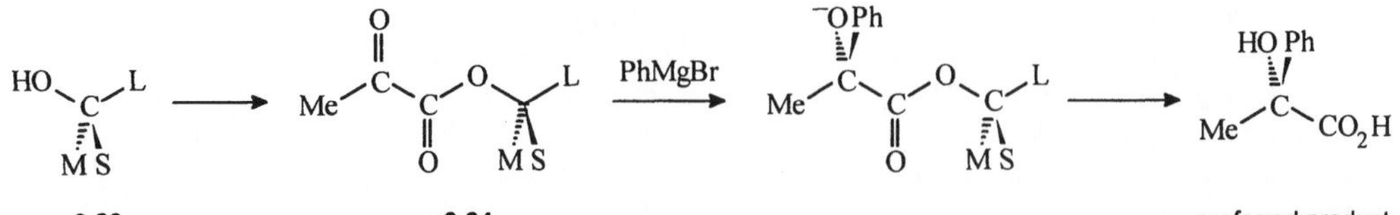

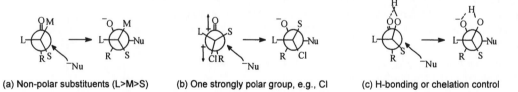

6.23 **6.24** preferred product

Fig. 6.51. An application of Prelog's Rule

A more convenient way of determining absolute configurations of chiral alcohols such as **6.23** relies on Horeau's Rule. The homochiral alcohol **6.23** reacts selectively with (*S*)-2-phenylbutanoyl groups in the racemic acid chloride or anhydride (**6.25**, used in excess), giving an excess of (*R*)-**6.26** after hydrolysis of the excess reagent (Fig. 6.52). An experimental advantage of this method compared with using Prelog's Rule is that it can be carried out reliably on a microscale and is sensitive enough to detect the difference in size between, e.g., CH_3 and CD_3 in $CH_3CHOHCD_3$ (CH_3 larger than CD_3).

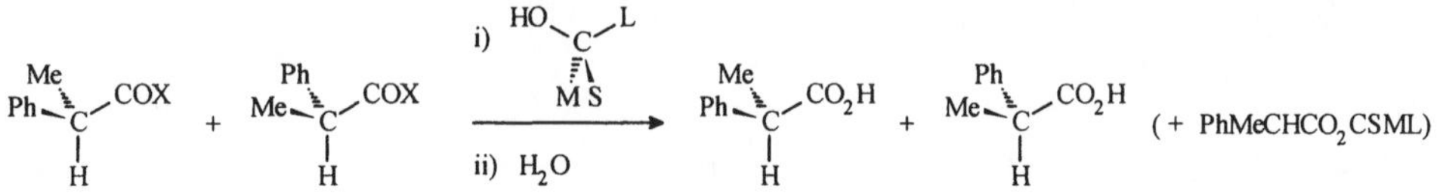

(*R*)-**6.25** (*S*)-**6.25** (*R*)-**6.26** (*S*)-**6.26** (+ PhMeCHCO$_2$CSML)
Racemic >50% <50%

Fig. 6.52. An application of Horeau's Rule [X = Cl or PhMeCHCO$_2$–]

Many of these reactions involve nucleophilic addition to a C=O group in a chiral aldehyde or ketone. Variants of *Cram's Rule* may be used to predict the diastereoselectivity in such reactions of *acyclic carbonyl compounds* with a chiral centre adjacent to the C=O group. If the three other groups are of low polarity and can be clearly ordered in size as large (L), medium (M), and small (S) then the major diastereomeric product may be predicted from the least hindered approach of an *achiral* nucleophilic reagent to the substrate in the conformations shown in Fig. 6.53(a). If one group is strongly polar, e.g., Cl (b), or if there is hydrogen bonding or chelation between one group and the O atom (c), then different reactant conformations must be assumed.

(a) Non-polar substituents (L>M>S) (b) One strongly polar group, e.g., Cl (c) H-bonding or chelation control

Fig. 6.53. Three variants of Cram's Rule for nucleophilic addition to C=O in acyclic ketones and aldehydes. The nucleophile is assumed to prefer to attack the less hindered side of the C=O in the conformations shown.

Cram's Rule and its variants have limitations. They assume that the preferred direction of attack is qualitatively independent of the nucleophile.

This does not hold for some cyclic ketones, notably derivatives of cyclohexanone (Fig. 6.54), even for simple nucleophiles. With chiral nucleophiles another factor comes into play, *double stereodifferentiation* (also called *double asymmetric synthesis or induction*).

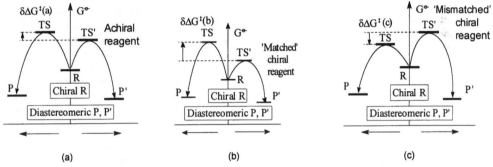

Fig. 6.54. Reduction of 4-t-butylcyclohexanone by a large (*eq* approach) and by a small (*ax* approach) reagent

6.9 Double stereodifferentiation (double asymmetric induction)

We have seen above that homochiral reagents (or catalysts) can give good *enantio*selectivity when achiral substrates are converted into enantiomeric compounds. If both the substrate *and* the reagent are chiral then the two may

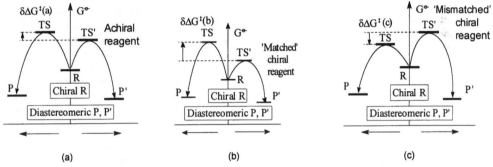

Fig. 6.55. In (a) $\delta\Delta G^{\ddagger}$(a) is a measure of the diastereoselectivity in the reaction at a prostereomeric centre of a chiral reactant R with an achiral reagent to generate a new chiral centre. This increases in (b) to $\delta\Delta G^{\ddagger}$(b) when the appropriate enantiomer of a chiral reagent is used, biasing the reaction even more towards P'. If the other enantiomer of the reagent is used the selectivities are 'mis-matched'. If the reagent selectivity is dominant, i.e., $\delta\Delta G^{\ddagger}$(c) is opposite in sign to $\Delta G^{\ddagger}$(a), the diastereomeric product P then predominates, as in (c).

act cooperatively ('matched') or antagonistically ('mismatched'). If the influence of the reagent is dominant then the stereoselectivity may be changed *qualitatively*, as in Fig. 6.55(c). Sharpless' Rule applied to the epoxidation of homochiral allylic alcohols (Fig. 6.43, page 79) is a good example of this. Diastereoselectivity in mis-matched reactions of chiral aldehydes with chiral enol boronates is often dominated by the enol component and then goes against Cram's Rule.

The chemistry of enol boronates is covered by S.E. Thomas, *Organic Synthesis: the Roles of Boron and Silicon*, Oxford University Press, Oxford, 1991.

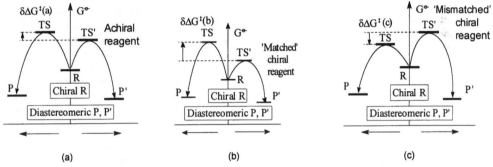

Fig. 6.56

Aldol reactions of **6.28** (Fig. 6.56 and 6.57) illustrate examples (a) to (c) in Fig. 6.55. When an achiral ketone **6.27** is used in reaction with (*R*)-**6.28** the new chiral centres are formed in the (S,S)-configuration with moderate selectivity (Fig. 6.56). When the homochiral ketone **6.29** is used (Fig. 6.57) (S)-**6.28**, but not (*R*)-**6.28**, gives high diastereoselectivity.

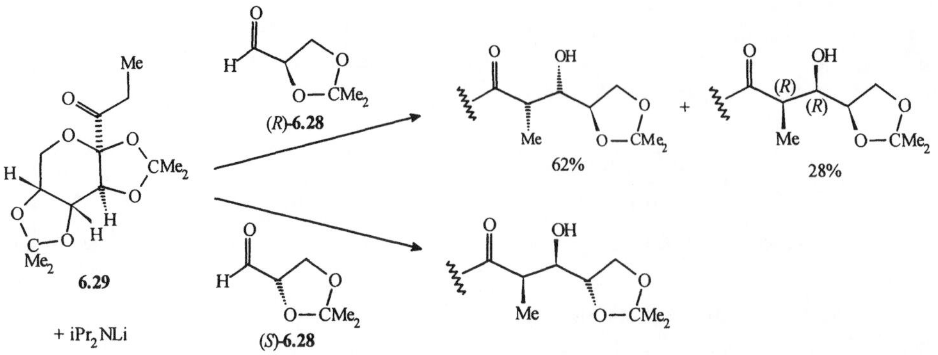

Fig. 6.57. Matched and mismatched aldol reactions between a homochiral enol and enantiomeric aldehydes

The use of chiral auxiliaries and catalysts

Sometimes available homochiral reagents show little enantio- or diastereo-selectivity, e.g., alkylating agents in S$_N$2 reactions, or are not available at all. Often we need to use achiral reagents, such as cyclopentadiene and acrylic acid in a Diels Alder reaction (Fig. 6.58), to make a chiral product. One way to solve this problem is to use a *chiral auxiliary* that can convert an achiral reactant into a homochiral derivative. After a diastereoselective Diels Alder reaction, with a Lewis acid catalyst, the chiral auxiliary is split off leaving a homochiral product of an overall enantioselective sequence. In effect we change from the situation in Fig. 6.38 to that in Fig. 6.50, so far as the critical step is concerned.

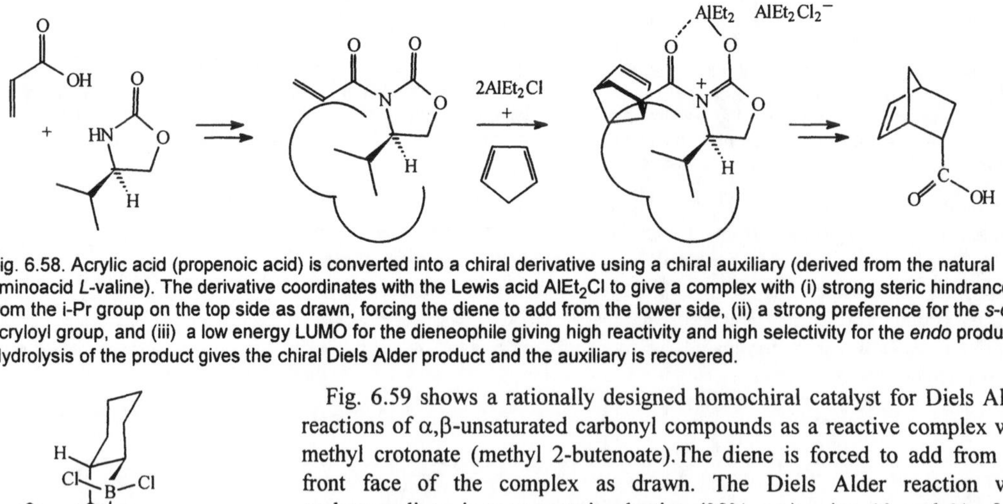

Fig. 6.58. Acrylic acid (propenoic acid) is converted into a chiral derivative using a chiral auxiliary (derived from the natural aminoacid L-valine). The derivative coordinates with the Lewis acid AlEt$_2$Cl to give a complex with (i) strong steric hindrance from the i-Pr group on the top side as drawn, forcing the diene to add from the lower side, (ii) a strong preference for the *s-cis* acryloyl group, and (iii) a low energy LUMO for the dienophile giving high reactivity and high selectivity for the *endo* product. Hydrolysis of the product gives the chiral Diels Alder product and the auxiliary is recovered.

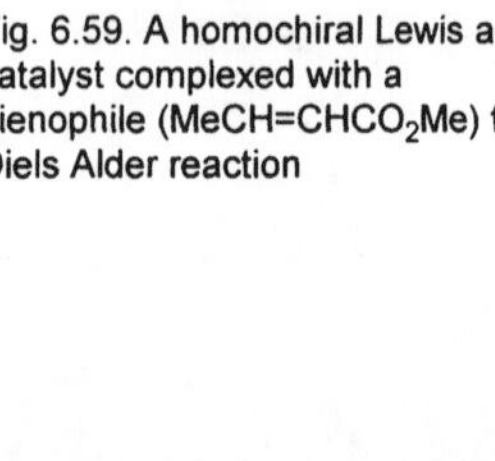

Fig. 6.59. A homochiral Lewis acid catalyst complexed with a dienophile (MeCH=CHCO$_2$Me) for a Diels Alder reaction

Fig. 6.59 shows a rationally designed homochiral catalyst for Diels Alder reactions of α,β-unsaturated carbonyl compounds as a reactive complex with methyl crotonate (methyl 2-butenoate).The diene is forced to add from the front face of the complex as drawn. The Diels Alder reaction with cyclopentadiene is very enantioselective (98% e.e.) using 10 mole% of the catalyst. Whenever such a catalyst is available it has a great advantage over multistep reactions, as in Fig. 6.58, that are typical of applications of chiral auxiliaries.

Problems

1. (a) How many diastereomers are there of 1,2,3,4,5,6-hexachlorocyclo-hexane? (b) Which, if any, are chiral? (c) Identify the diastereomer which is ~25,000 slower in E2 reactions than the others.

2. Why is $-CH_3$ effectively larger than $-CD_3$ in many reactions (see, e.g., Horeau's Rule, p. 82)?

3. Devise molecules and reactions such that (i) a reaction with inversion at a chiral centre does *not* change the descriptor R for that centre, and (ii) a reaction with retention at a chiral centre changes the descriptor from R to S.

4. Identify the diastereomers of the products that result from the following stereospecific transformations:

(a) E-Pr–CH=CH–SiMe$_3$ → Pr–CH=CH–Pr

   Reagents: (i) PhCOO$_2$H; (ii) Pr$_2$CuLi; (iii) NaH

(b)   Bu–C≡C–H → Bu–CH=CH–(c-C$_6$H$_{11}$)

   Reagents: (i) HB(c-C$_6$H$_{11}$)$_2$; (ii) NaOH/H$_2$O/I$_2$ [c-C$_6$H$_{11}$ = cyclohexyl]

(c)   Bu–C≡C–Br → Bu–CH=CH–(c-C$_6$H$_{11}$)

   Reagents: (i) HB(c-C$_6$H$_{11}$)$_2$; (ii) NaOMe; (iii) MeCO$_2$H [c-C$_6$H$_{11}$ = cyclohexyl]

Further Reading

I. Fleming, *Frontier Molecular Orbital Theory and Organic Chemical Reactions*, Wiley, Chichester, 1976.

L.M. Harwood, *Polar Rearrangements*, Oxford University Press, Oxford, 1992.

L.N. Mander, *Stereoselective Synthesis*, in *Stereochemistry of Organic Compounds*, by E.L. Eliel and S.H. Wilen, Wiley, New York, 1994, pp. 835-990.

T.L. Gilchrist and R.C. Storr, *Organic Reactions and Orbital Symmetry* (2nd ed.), Cambridge University Press, London, 1979

W.H. Saunders and A.F. Cockerill, *Mechanisms of Elimination Reactions*, Wiley, New York, 1973.

S. Masamune, W. Choy, J.S. Peterson, and L.R. Sita, *Angew. Chem. Int. Ed. Engl.*, 1985, **24**, 1 ('Double asymmetric synthesis').

Rearrangements in Ground and Excited States, ed. P. de Mayo, Academic, New York, 1990.

Further Reading: General references

E.L. Eliel and S.H. Wilen, *Stereochemistry of Organic Compounds*, Wiley, New York, 1994.

K. Mislow, *Introduction to Stereochemistry*, Benjamin, New York, 1966 (out of date for nomenclature but very accurate: many good problems).

G. Natta and M. Farina, *Stereochemistry*, Longman, London, 1972: easy to read introduction, with good diagrams, and interesting chapters on polymers.

Problems: hints and answers

Chapter 2

1. (a) C_s; (b) $C_{\infty v}$; (c) D_{2h}; (d) D_{2d}; (e) D_2; (f) S_4; (g) D_n; (h) S_n; (i) C_2; (j) C_{4v}; (k) I_h; (l) Variable: see I. Stewart, *Scientific American*, February 1997, pp. 80-81.

2. (a) C_2; (b) (i) C_{3v}, (ii) C_{2v}, (iii) C_{2v}, (iv) D_{3d}; (c) *sc*: C_2, *ap*: C_{2h}; (d) C_2; (e) D_2; (f) C_2; (g) *Z*: C_{2v}; *E*: C_{2h}; (h) C_2.

3. (a) Yes; (b) (i), (ii), (iii), and (v): yes, (iv): no; (c) no. See "La coupe du roi", *J. Am. Chem. Soc.*, 1982, **104**, 1419-1426, and the example in the margin.

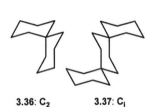

Two *identical chiral* half cubes: one of an infinite number of possible cuts.

Chapter 3

1. (a) (i) Slow rotation about N–CO at low temperatures, fast at high (*Me* signals coalesce). (ii) to (iv) Slow chair to chair ring inversion at low temperatures, fast at high.

(b) (i) 1 (*C*=O); (ii) 1 (for C-9,10); (iii) 0; (iv) 2 (for C-2 and C-5); (v) 0.

(c) (ii) C_2, C_{2v}; (iii) C_1, C_s; (iv) C_1, C_2; (v) C_s, C_{2v}.

2. (a) See **3.36** in margin; (b) C_2; (c) (i) Changes into enantiomer; (ii) No change.

3. (a) See **3.37** in the margin; (b) C_i; (c) You get (i) a chiral conformer (C_2), in which at most two rings can be viewed as chairs at the same time, (ii) no change.

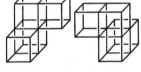

3.36: C_2 **3.37**: C_i

4. 11 (3 meso, 4 pairs of enantiomers).

Chapter 4

1. **4.11**: *R*; **4.12**: *S*; **4.13**: 1*R*,2*R*,3*S*,4*R*,P-*R*; **4.14**: *S*; **4.21**: *S*,*S*; **4.37**: 1*R*,3*S*,4*S*; **4.39**: 3*S*,8*S*,9*S*,10*S*,14*S*,17*R*, 20*S*; Fig. 4.11: *S*, *S*, *R*, *R*.

2. **4.25**: *M*; **4.26**: *P*; **4.27**: *P*.

3. H_a: *pro-S*; H_b: *pro-R*; $(CO_2H)_c$: *pro-S*; $(CH_2CO_2H)_d$: *pro-S*. H_a: *Si*; H_b: *Re*; $(CO_2H)_c$: *Si*; $(CH_2CO_2H)_d$: *Si*. N.B. In this instance there is a simple correspondence between *pro*-R/S and *Re/Si* but this is not general: see page 57.

4. *lll*, D_2; *luu*, S_4; *lul*, C_2; *uuu*, C_1.

5. **4.66**: 3 meso and 2 pairs of enantiomers. **4.67**: 2 meso and 3 pairs of enantiomers.

6. See structure **A** below: only cyclopentane-*cis*-1,2-dicarboxylic acids form cyclic anhydrides.

7. (a) **4.68**: 2 meso (*cis-cisoid-cis* and *cis-transoid-cis*) and 4 pairs of enantiomers; **4.69**: 3 meso (*cis-cisoid-cis*, *trans-cisoid-trans*, and *cis-transoid-cis*) and 2 pairs of enantiomers. (b) the (meso) *trans-cisoid-trans* diastereomer of **4.68** (**B**) and the (chiral) *trans-transoid-trans* diastereomer of **4.69** (**C**, one enantiomer) must have the central rings in (twist) boat conformations.

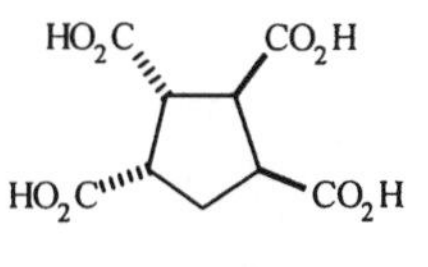

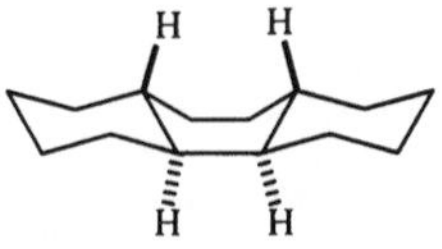

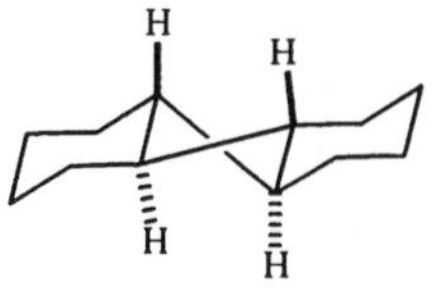

A   **B**   **C**

8. (a) The Hs in **4.70** are enantiotopic down to 185K (rapid N inversion giving an effective plane of symmetry through N and all three attached Cs) but diastereotopic at and below 122K (slow N inversion: symmetry C_2: σ through N and Me only).

In acidic aqueous solutions most molecules of **4.70** are protonated and cannot directly invert but are in mobile equilibrium with the small amount of free base which can (see Fig. 4.9). As the pH is lowered the proportion of free base decreases and therefore the rate at which the cation can invert at the nitrogen atom decreases, mirroring the effect of lower temperature in the free base.

(b) Cycloocta-1,3,5,7–tetraene is non-planar but achiral, with symmetry D_{2d} (with two planes of symmetry), and there is a substantial energy barrier (~62 kJ mol^{-1}) to ring inversion through a planar conformational transition state. In 1,2,3,4–tetramethylcycloocta-1,3,5,7–tetraene (**D**) the methyl groups reduce the symmetry to C_1. The methyl groups also raise the barrier to ring inversion because the 2- and 3-methyl groups are very close in the transition state.

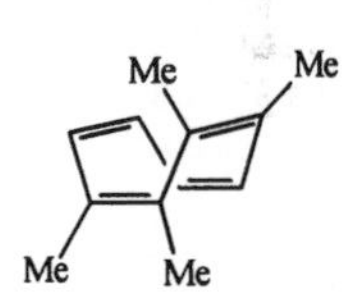

D

Chapter 6

1. (a) 8; (b) 1 (**E**) and its enantiomer; (c) **F** (E2 must involve *cis*-H, Cl)

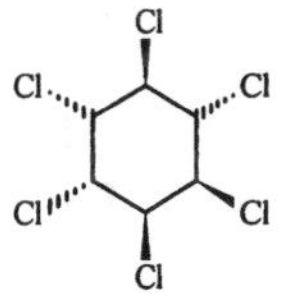

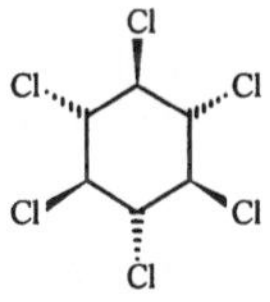

3. See, e.g., A. Horeau, A. Nouaille, and K. Mislow, *J. Am. Chem. Soc.*, 1965, **87**, 4957.

3. (i) S_N2 reaction of **G** with $^-C\equiv N$.   (ii) Reaction of **H** with MeCO$_2$H.

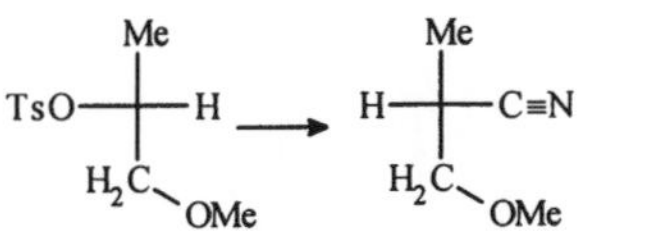

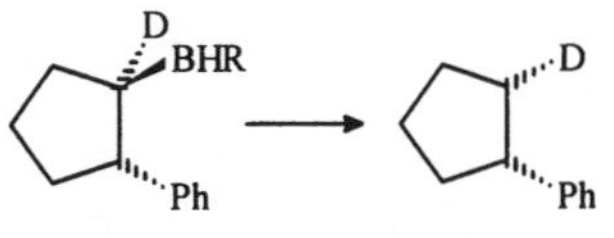

E   **F**

G   **H**

The homochiral borane **H** could be prepared from 1-phenylcyclopentene-2-d, borane, and (–)-α-pinene (see **I** in margin), via (–)-isocampheylborane (**J**).

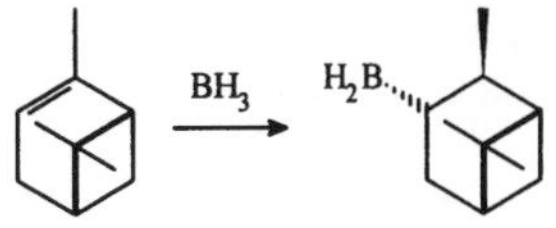

I   **J**

4. (a) *E*-Pr–CH=CH–Pr; (b) *Z*-Bu–CH=CH–(c-C$_6$H$_{11}$);
(c) *E*-Bu–CH=CH–(c-C$_6$H$_{11}$).

For more information related to problem 4(ii) see S.E. Thomas, *Organic Synthesis: The Roles of Boron and Silicon*, Oxford University Press, Oxford, 1991, pp., 11, 68-69, 25, 26.

INDEX